HUMAN RESPONSE TO VIBRATION

HUMAN RESPONSE TO VIBRATION

Neil J. Mansfield

CRC PRESS

Boca Raton   London   New York   Washington, D.C.

Library of Congress Cataloging-in-Publication Data

Mansfield, Neil J.
　　Human response to vibration / Neil J. Mansfield.
　　　　p.　cm.
　　Includes bibliographical references and index.
　　ISBN 0-415-28239-X
　　1. Vibration—Physiological effect. I. Title.

QP82.2.V5M36 2004
612′.01445—dc22   2004054491

Visit the CRC Press Web site at www.crcpress.com

© 2005 by CRC Press LLC

No claim to original U.S. Government works
International Standard Book Number 0-415-28239-X
Library of Congress Card Number 2004054491
Printed in the United States of America　1　2　3　4　5　6　7　8　9　0
Printed on acid-free paper

Preface

I feel fortunate to work in the field of environmental ergonomics, with a particular interest in human response to vibration. Those working in this discipline have proved, over many years, to be a welcoming group of people, and many of my former colleagues and collaborators I now regard as friends. Through continued collaboration and sharing of ideas, data, and results, this international community of researchers and practitioners has developed an understanding of the many facets of human response to vibration; however, we still have a long way to go, and there are still many unknowns.

This book has been written with the consultant, practitioner, and student in mind, although it is also hoped that the research community will find it helpful. It is designed to educate, to be used for reference and, hopefully, to stimulate new ideas for the next generation of specialists. In many areas of human response to vibration we have confidence in our understanding; in many other areas there is conflicting opinion and data. Thus, one must approach the topic with caution. It is hoped, and expected, that some of the ideas presented in this book will develop as time passes, and therefore this publication can only represent a snapshot of understanding, rather than a definition of fact. This is the nature of scientific research and one of the attractive aspects of the profession.

Unfortunately, no book can osmotically transfer experience, and so readers must be prepared to get on site and to gain their own experiences. Fortunately, in many cases, time on site can be spent working on interesting projects, in interesting places, and with interesting people. These experiences can be quite enjoyable.

Usually, human vibration specialists aim to reduce the exposure to, and the effects of, the mechanical stimulus. Indeed, the concept of excessive vibration being unpleasant is well established in the vernacular and has been for centuries. Although the Beach Boys might have been pleased to experience their "Good Vibrations," my own foundation is well summarized by the Psalmist King David:

My soul finds rest in God alone; my salvation comes from him. He alone is my rock and my salvation; he is my fortress, I will never be shaken.

Psalm 62:1–2 New International Version

Neil J. Mansfield

Acknowledgments

I would like to thank several groups of people who have made this book possible.

First, I must thank Gurmail Paddan, Rupert Kipping, and Sharon Holmes for their helpful comments on the first drafts of the manuscript. Their technical, medical, and literary advice was valuable.

I am grateful to former colleagues for providing inspiration in the field, especially Lage Burström, Mike Griffin, Patrik Holmlund, Chris Lewis, Ronnie Lundström, and Setsuo Maeda, in addition to all of the members of the "UK Group." Colleagues within the Environmental Ergonomics Research Cluster of the Department of Human Sciences, Loughborough University, make my research a pleasure to undertake, and I would like to acknowledge Ken Parsons, George Havenith, Peter Howarth, Andrew Rimell, Simon Hodder, Sarah Atkinson, Stacy Clemes, Aili Haasnoot, Gerry Newell, and Luca Notini for their contributions.

It would not have been possible to complete this book without the unwavering support of Caroline, Jack, Finn, and Lotta. I am fortunate to have such a wonderful family.

Contents

1 Introduction to Vibration

1.1 HUMAN RESPONSE TO VIBRATION IN CONTEXT

Humankind has always had the desire to build, create, and explore. Each of these activities has involved exposure to vibration, whether the source comes from primitive axes, handsaws or riding in carts, from using power tools, industrial machines, or riding in planes, trains, and automobiles. As modern man has harnessed power sources in more efficient ways, the apparatus used to build, create, and explore have used more energy and, as a result, increased quantities of energy have been dissipated in the form of vibration, some of which has been transmitted to people.

One of the most rewarding aspects of studying human response to vibration is its truly multidisciplinary nature (Figure 1.1). For example, the authors listed in the references have varied backgrounds that include engineering, psychology, the natural sciences, clinical medicine, and ergonomics. If the phrase "human response to vibration" is deconstructed into its component parts, then we can consider that a complete grasp of the discipline requires an understanding of the human (biological, anatomical, and physiological aspects), their response (psychological and biomechanical aspects), and the nature of the vibration (in terms of the engineering and underlying physics). Although most researchers have a focus in one of the three main areas, it is the understanding of the interactions between these component parts that is essential in developing our understanding of the topic.

The human aspects and response aspects are the focus of Chapter 2, Chapter 3, and Chapter 4 of this book. Other chapters consider the practicalities of applying this knowledge and the frameworks that are defined in standards and directives. This chapter considers the underlying nature of vibration itself and defines some of the vocabulary that will be used throughout the rest of the book. It considers the fundamentals of wave theory that underpin the discipline (Section 1.2), how vibration is classified when it comes to human response (Section 1.3), and how vibration axes are defined (Section 1.4).

1.2 INTRODUCTION TO WAVE THEORY

Vibration is mechanical movement that oscillates about a fixed (often a reference) point. It is a form of mechanical wave and, like all waves, it transfers energy but not matter. Vibration needs a mechanical structure through which to travel. This structure might be part of a machine, vehicle, tool, or even a person, but if a mechanical coupling is lost, then the vibration will no longer propagate.

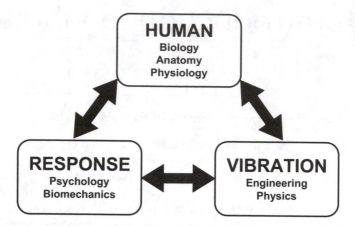

FIGURE 1.1 Component parts and topic areas for the discipline of "human response to vibration."

This section gives only an introduction to the fundamentals of wave theory. For a more complete discussion of the topic, the reader could consult, for example, Kinsler et al. (2000) or Harris and Piersol (2002).

1.2.1 Simple Waves

The most simple type of wave is defined mathematically as:

$$a(t) = A \, sin \, (2\pi f t)$$

where $a(t)$ is the acceleration (measured in m/s^2) at time t. This wave has an amplitude A and a frequency f cycles per second (unit = hertz, Hz; Figure 1.2). Such waves are known as sine waves. If the frequency of the wave increases, then the period of the wave will decrease. This means that as each individual cycle takes less time, the frequency rises. In the engineering literature, frequency can also be expressed in terms of radians per second (ω) where:

$$\omega = 2\pi f$$

To define a simple sine wave, we need to know its frequency, its amplitude, and the time from the starting point. Unfortunately, human vibration exposures are rarely these simple sine waves. Therefore, more complex descriptions of waves are required.

1.2.2 Adding Waves

Complex waves can be produced by the addition of sine waves with different amplitudes, frequencies, and phases (i.e., time delays from the start of the wave). This principle of superposition means that when individual component waves interact, the resultant vibration is equal to the sum of all of the components. For example, if five waves of different frequencies are summed, then a complex wave results

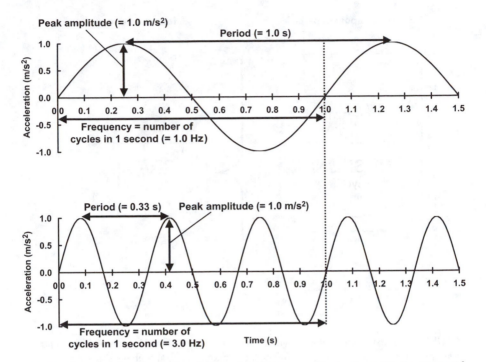

FIGURE 1.2 Basic descriptors for a 1-Hz and a 3-Hz wave with a peak amplitude of 1.0 m/s².

(Figure 1.3). The beauty of this phenomenon is that the superposition process can be reversed using mathematical techniques such that a complex wave can be resolved into simple component parts, each with different amplitudes, phases, and frequencies. It is this process that allows for consideration of human responses to vibration with simultaneous components of different frequencies. These are the types of vibration signals that are encountered in the real world (e.g., when driving a car, the occupants are simultaneously exposed to low-frequency vibration caused by the general road profile in addition to higher-frequency vibration caused by the roughness of the road surface).

1.2.3 DISPLACEMENT, VELOCITY, AND ACCELERATION

Any vibration signal has three qualities: its displacement, velocity, and acceleration, which are inextricably linked. Consider a slow-moving vertical oscillation, such as a ship rising and falling on large waves. The ship rises on each wave, stops on the wave crest, falls into the trough, and stops and rises again on the next wave. The maximum vertical displacement occurs when the ship is at the top of each wave, yet this coincides with the instant of zero velocity. The greatest velocity occurs while rising or falling (positive or negative). The minimum vertical displacement occurs when the ship is in each trough, when, again, there is zero velocity. There is also an associated cyclic acceleration and deceleration due to the constantly changing velocity. For any wave, the displacement, velocity, and acceleration do not coincide

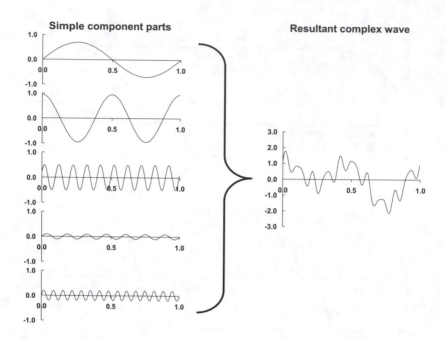

FIGURE 1.3 Illustration of how a complex wave can be produced by adding a small number of simple sine waves with different frequencies, magnitudes, and phase shifts.

with one another; indeed, for a sine wave, the displacement and acceleration have an inverse relationship (Figure 1.4).

Depending on the frequency of the vibration, the displacement, velocity, or acceleration can dominate the mechanism of perceiving whether an object is vibrating. At low frequencies, displacement is the most important property. For example, a person looking out of the window from the top floor of a very tall building on a blustery day might be able to see that he or she is slowly swaying (as tall buildings are designed to do) by observing their relative displacement to the ground, yet not perceive any acceleration or velocity. At intermediate and high frequencies, velocity and acceleration are the most important properties, respectively. For example, it is often easy to perceive the high-frequency vibration from a fan in a piece of electrical equipment by lightly touching the surface of the casing, yet there is no other cue that any vibration exists (e.g., visual).

1.2.4 RESONANCE

If a compliant mechanical structure is oscillated very slowly, then it will move as a single coherent unit, acting as a pure mass. However, at high frequencies, the vibration can be localized to the point of application, i.e., the structure is isolated from the vibration. It is this principle that is used in car suspensions, engine mounts, and in isolation mechanisms for turntables and CD players. Between these high and low frequencies is a zone where the response of the system will be maximized when compared to the stimulus (Figure 1.5). This is known as reso-

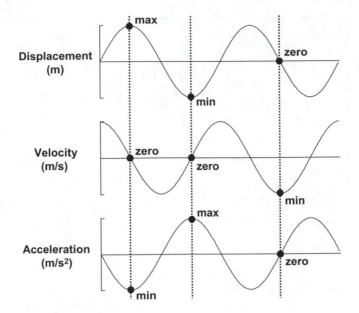

FIGURE 1.4 Displacement, velocity, and acceleration for a sine wave showing asynchronous peaks and troughs.

nance. All compliant systems have a resonance frequency, and complex structures have more than one. Many famous disasters have ultimately resulted from a resonant failure, including the 1940 Tacoma Narrows bridge collapse where a new suspension bridge was unable to withstand the high winds that induced vibration at its resonance frequency.

To ensure that vibration at the resonance frequency does not build up to a point of failure, engineering structures are damped. However, there is a trade-off, as the more damping that exists in a system, the less effective it will be at isolating from high frequency vibration. Humans are inherently highly damped, although resonances are still clearly observable. These resonances mean that if an individual is exposed to vibration, his or her response will depend not only on the magnitude but also on the frequency of the stimulus.

1.2.5 REPRESENTATION OF VIBRATION SIGNALS

Vibration can be presented in a graphical form where the x-axis of the graph represents time and the y-axis represents the acceleration at any time. This is known as representation in the "time domain." Although time-domain representations are useful in understanding the waveform of the motion, they are difficult to interpret and cannot usually be applied to standardized analysis methods. Time-domain signals are always presented with linear axes.

An alternative representation of vibration is to present it in the "frequency domain" where the x-axis of the graph is frequency and the y-axis is vibration magnitude. This allows for a representation of how much vibration energy exists

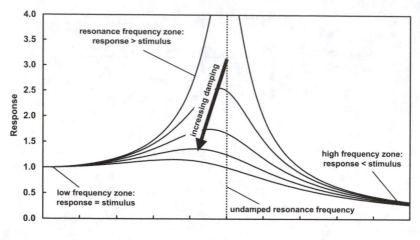

FIGURE 1.5 Response of a simple dynamic system to vibration. At low frequencies, the response equals the stimulus; around the resonance frequency the response is greater than the stimulus; at high frequencies the response is less than the stimulus. As damping increases, the peak response decreases.

at each frequency and is often more useful than time-domain representations for practical purposes. Depending on the application, frequency-domain graphs can be presented with linear or logarithmic axes. For whole-body vibration, the x-axis is usually linear; for hand-transmitted vibration, the x-axis is usually logarithmic. Although frequency-domain graphs are sometimes presented with linear y-axes, human responses to physical stimuli are usually proportional to the logarithm of the stimulus magnitude (e.g., Stevens' Law; Stevens, 1957) and so a logarithmic y-axis is usually most appropriate when considering the human response to the vibration.

1.3 CLASSIFICATION OF VIBRATION

Vibration can be classified using a variety of descriptors. Some of these terms are technical and have specific meaning; others are more generic and the meaning is dependent on the context of use.

1.3.1 CLASSIFICATION OF VIBRATION BY CONTACT SITE, EFFECT, AND FREQUENCY

People are primarily exposed to either localized vibration or vibration that affects the whole body. Usually, localized vibration only affects the hand-arm system, as it is caused by an individual holding a vibrating object such as a tool, workpiece, or control device (e.g., a joystick, handlebars, or steering wheel). In this case, the vibration is termed "hand-transmitted" or "hand-arm" vibration. These terms are

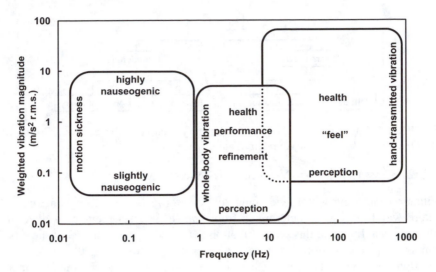

FIGURE 1.6 Typical frequency ranges and magnitudes of interest for the study of motion sickness, whole-body vibration, and hand-transmitted vibration.

sometimes abbreviated to HTV or HAV, respectively. If this vibration is of sufficient magnitude and occurs long enough, then hand-arm vibration syndrome (HAVS) can develop. HAVS is characterized by disorders of the muscles, nerves, bone, joints, and circulatory systems, the most well known of which is vibration-induced white finger (VWF). The effects of hand-transmitted vibration are most marked at relatively high frequencies, with the range of 8 to 1000 Hz generally considered to be the most important (Figure 1.6). Localized vibration can also occur at other sites, such as at the feet due to pedal vibration, but it is unusual to observe any adverse human responses in these situations.

"Whole-body vibration" is vibration that affects the whole of the exposed person, i.e., the vibration affects all parts of the body. It is usually transmitted through seat surfaces, backrests, and through the floor, although it can also be important for those standing or lying, such as for those traveling on busy commuter trains or patients being transported by ambulance. Most whole-body vibration exposures are associated with transportation where vehicle drivers or passengers are exposed to mechanical disturbances and impacts while traveling. Whole-body vibration can affect comfort, performance, and health, depending on the magnitude, waveform, and exposure times. People are most sensitive to whole-body vibration within the frequency range of 1 to 20 Hz, although many measurements include higher frequencies. Whole-body vibration is sometimes abbreviated to WBV.

The final classification of human vibration by contact site, effect, and frequency is vibration that causes motion sickness. Motion sickness can occur when a person is exposed to real or apparent low-frequency motion (below 1 Hz). Although it is true that the whole body is affected by such stimuli, the frequency range of interest and the effects of the vibration are distinct from those relevant to whole-body vibration and so the topic is usually considered separately.

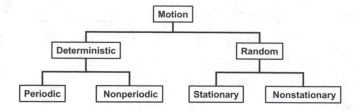

FIGURE 1.7 Categorization of types of oscillatory motion. [Adapted from Griffin, M.J. (1990). *Handbook of Human Vibration.* London: Academic Press.]

1.3.2 CLASSIFICATION OF VIBRATION BY WAVEFORM

Vibration can occur with a wide range of waveforms (Figure 1.7). Sometimes the future waveform can be determined from knowledge of a previous waveform or mathematical function; this type of motion is termed "deterministic." If the motion cannot be predicted from previous events or other knowledge, then it is termed "random." Deterministic motion can be of two types. Either the vibration is a motion that is repeated, whereby each successive cycle is identical ("periodic" motion, such as a sine wave), or the vibration could only occur as one cycle at a time ("nonperiodic" motion). Nonperiodic motion is sometimes subcategorized as "transient" or "shock" motion.

In practice, vibration that humans are exposed to is usually random. If the random motion's statistical properties do not change over time, then the vibration is said to be "stationary." If the statistical properties do change over time, then the vibration is said to be "nonstationary." When vibration is measured in an environment, it is usually assumed that the vibration is stationary. This means that the measurement that is being taken is representative of any other measurement that could be taken (i.e., the statistical properties would not change over time). However, this assumption is never perfectly true and is, occasionally, wildly inaccurate. When interpreting or generating measurement data, the nonstationarity should be considered, especially if the results are close to critical values (e.g., when ranking the "best" and "worst" machine or assessing compliance to a standard or legal obligation).

1.3.3 CLASSIFICATION OF VIBRATION BY MAGNITUDE

Different industrial sectors treat vibration in different ways, essentially due to associated magnitudes and coupled effects. At low magnitudes, issues of refinement and perception can be important (Figure 1.6). These issues form the focus for research and product development in sectors such as the automotive industry, where seat and steering wheel vibration is tuned to optimize comfort and perceptions of quality. Sports equipment manufacturers might be concerned with improving the "feel" of their products by optimizing the transmission of vibration to the competitor. These industries have a focus on the psychophysical aspects of the human response to vibration and need not be concerned with, for example, health effects.

At higher magnitudes, vibration could cause discomfort, reduce performance and induce activity interference. This is of interest to, for example, the military

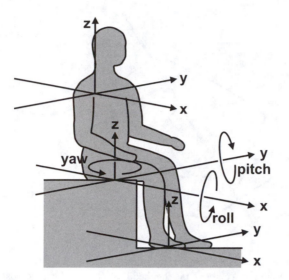

FIGURE 1.8 Vibration axes for whole-body vibration centered at the feet, seat, and back. [Adapted from British Standards Institution (1987). Measurement and evaluation of human exposure to whole-body mechanical vibration and repeated shock. BS 6841. London: British Standards Institution.]

sector where individuals are required to perform optimally under potentially challenging environmental conditions. In this context, considerations of "feel" or "ride quality" are not relevant.

For those exposed to relatively high magnitudes of vibration for extended periods of time (or extreme magnitudes for short periods of time), injury could occur. This is primarily of interest in an occupational health context. Manufacturers of machines, vehicles, or tools have an interest in reducing the vibration to give a competitive edge, whereas purchasers seek to protect their workforce with affordable, yet optimized, products with an acceptable magnitude of vibration emission. Again, these stakeholders are not concerned with the refinement of the dynamic environment.

1.4 VIBRATION AXES

Vibration can occur in any direction. Complex stimuli simultaneously move vertically, laterally, and in the fore-and-aft directions. In addition, rotation is possible, producing a total of six axes of potential movement.

For whole-body vibration, the fore-and-aft direction is defined as the x-axis, lateral as the y-axis, and vertical as the z-axis (Figure 1.8). Roll is rotation around the x-axis, pitch is rotation around the y-axis, and yaw is rotation around the z-axis. Therefore, considering head motion, roll corresponds to tilting the head to either side, pitch corresponds to a nodding motion, and yaw corresponds to the conventional shaking of the head. For a standing person, the axes are those for the feet (i.e., while sitting, the thighs are aligned to the x-axis, but are aligned to the z-axis when standing). For a supine person, the axes, relative to the individual, are the same as

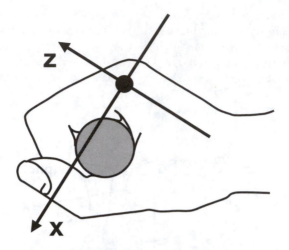

FIGURE 1.9 Vibration axes for hand-transmitted vibration. The y-axis is out of the page. The coordinate system is centered on the head of the third metacarpal. [Adapted from International Organization for Standardization (1997). Mechanical vibration and shock: Human exposure — biodynamic coordinate systems. ISO 8727. Geneva: International Organization for Standardization.]

for standing (i.e., foot-to-head is the z-axis). It should be noted that other specialisms within biomechanics use different coordinate systems to those that are used for whole-body vibration (e.g., gait analysis). Therefore, to avoid confusion, it is usually appropriate to define the x-, y-, and z-axes within any document.

For hand-transmitted vibration, there is a single coordinate system that is based on the head of the third metacarpal (i.e., the middle knuckle). The x-axis is through the palm, the y-axis is across the palm towards the thumb, and the z-axis extends towards the fingers parallel with the back of the hand (Figure 1.9). Defining the coordinate system at the third metacarpal ensures that there is no confusion regarding axes for palm or power grips or any orientation of the hand. Although it is probable that rotational vibration will affect the hand, it has rarely been studied, and no standardized terminology exists other than rotation about the x-, y-, or z-axes.

1.5 CHAPTER SUMMARY

Human response to vibration is a multidisciplinary topic involving biology, psychology, biomechanics, and engineering. Although these topics are sometimes considered in isolation, understanding the interactions between them is key to gaining a grasp of the topic area. Vibration is mechanical oscillation about a fixed reference point. In practice, people are exposed to complex waveforms, although these can be represented by their component parts defined as a summation of sine waves. Human vibration is usually classified as either hand-transmitted vibration, whole-body vibration, or motion sickness, and these subdisciplines are often considered separately from one another. Motion sickness is only concerned with frequencies below 1 Hz, whole-body vibration is concerned with frequencies from about 1 to 100 Hz, and hand-transmitted vibration is concerned with frequencies from about 8 to 1000 Hz. Vibration occurs in six axes simultaneously. For whole-body vibration, these are the x-axis, y-axis, and z-axis (fore-and-aft, lateral, and vertical, respectively) and rotation about these axes (roll, pitch, and yaw). Often, rotational vibration is neglected in vibration analysis. For hand-transmitted vibration the x-axis passes through the palm, the y-axis across the palm, and the z-axis parallel with the back of the hand in the direction of the fingers.

2 Whole-Body Vibration

2.1 INTRODUCTION

It is common for people to experience whole-body vibration on most days of their lives. An office worker may travel to work in a car, a bus, a train, or use a bicycle or motorcycle. A factory worker may use industrial trucks; an agricultural worker may drive a tractor; military personnel travel in many types of tracked and wheeled vehicles across rough terrain, or may fly in aircraft or travel in ships and boats. Other workers such as astronauts, jockeys, race car drivers, and cabin crew may be exposed to whole-body vibration or they may be paid to participate in experimental research studying the effects of vibration on humans. In these and many other environments, people are primarily exposed to vibration while seated and are exposed to a wide range of vibration magnitudes, waveforms, and durations.

Whole-body vibration occurs when a human is supported by a surface that is shaking and the vibration affects body parts remote from the site of exposure. For example, when a forklift truck drives over a bumpy surface, vibration is transmitted through the vehicle to the seat and footrest, which are the surfaces that support the driver. The vibration is then transmitted through the body of the driver to the head, which will move. This transmission path includes the seat (Section 2.5); the surface of the driver in contact with the vehicle (Section 2.2) including the driver's nervous system (Section 2.3); the skeleton, including the spine where an injury might occur (Section 2.4); and ultimately the skull, which might have its own dynamic responses to the transmitted vibration (Section 2.6).

2.2 WHOLE-BODY VIBRATION PERCEPTION

One could argue that vibration below human perception thresholds is unimportant, although vulcanologists and seismologists might disagree. The question therefore arises: What is the magnitude of vibration required for it to be perceived? It has been repeatedly shown that under laboratory conditions, perception thresholds are a function of frequency and direction. However, the psychological response to vibration exposure above threshold is also a function of the situation and type of exposure. For example, if the movement of the floor of an underground train were to be transferred to the floor of a building, one would expect a somewhat different response from those standing on that surface! So, considering the relative importance of exposure to earthquake vibration that might only be just above perception thresholds, perhaps our vulcanologists and seismologists should continue monitoring vibration that cannot be felt.

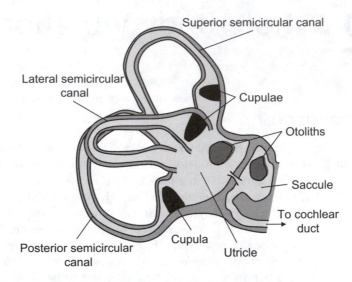

FIGURE 2.1 Anatomy of the vestibular complex, located in the inner ear.

2.2.1 MECHANISMS OF PERCEPTION

The body does not contain a single vibration-sensing organ but combines signals from the visual, vestibular, somatic, and auditory systems. Each of these systems can sense vibration in more than one way, and will be considered in turn.

For high-displacement and therefore low-frequency oscillation, one can clearly see movement by changes in the relative position of objects on the retina. The visual system might also sense vibration by observing motion of other objects in the vibration environment. For example, a rear-view mirror of a car might vibrate, blurring the image; drapes and lights might swing in response to movement; the surface of a drink might show ripples. Finally, the eyeball itself can resonate at frequencies between 30 to 80 Hz, blurring vision (e.g., Stott et al., 1993).

The vestibular complex of the inner ear includes the semicircular canals and the vestibule, which are sensitive to rotational and linear acceleration, respectively (Figure 2.1). Endolymph in the three orthogonally orientated semicircular canals has inertia. Hence, when the head rotates, the endolymph applies a force on the gelatinous cupula, which bends and stimulates the embedded hair cells and thus the nerves to which they are attached. The vestibule is composed of two sacs, the utricle and the saccule. Hair cells within these sacs are embedded in a gelatinous substance that is topped with small calcium crystals (otoliths). When the head is exposed to linear acceleration, or when it changes its orientation to gravity, the mass of the otoliths causes a distortion of the hair cells, which again are connected to the nervous system.

The somatic system can be subdivided into three elements: kinesthetic, visceral, and cutaneous. Kinesthetic sensation uses signals from proprioceptors in the joints, muscles, and tendons to provide feedback to the brain on the position and forces within segments. Similarly, visceral sensation uses receptors in the abdomen. Cutaneous sensation consists of a combined response of four types of nerve endings in the skin (see also Section 4.2). Ruffini endings are located in the deep layer of the skin (the dermis) and respond not only to high-frequency vibration (100 to 500 Hz) but also lateral stretching and pressure. Pacinian corpuscles are also located in the dermis but respond to vibration in the 40 to 400 Hz frequency range. Merkel's disks are located closer to the surface of the skin (the epidermis) and respond to perpendicular pressure at frequencies below 5 Hz. Meissner's corpuscles are also located close to the skin surface and are sensitive to vibration between 5 and 60 Hz.

The final sensory system for whole-body vibration is the auditory system. In most vehicles, exposure to transients and shocks can be heard by radiation through the structure of the vehicle. At higher frequencies (above 20 Hz), vibrating surfaces can also act as loudspeakers, directly driving the air, and this can result in auditory perception. Also, there is some evidence that skull-conducted vibration can be "heard" at approximately one tenth of the magnitude required for cutaneous feeling (Griffin, 1990).

This combination of sensory signals must be assimilated by the brain to produce a cognitive model of the motion environment. Therefore, psychophysical techniques are appropriate for the investigation of the human perception of vibration (e.g., Coren et al., 1999). Experimental studies have shown the maintenance of Stevens' Law, which states that sensation magnitude increases proportionally to the stimulus magnitude raised to some power:

$$S = cI^m$$

where S is the sensation magnitude, c is a constant, I is the stimulus magnitude, and m is the value of the exponent (see also Subsection 2.3.3).

Similarly, experimental studies have shown the maintenance of Weber's Law, which states that the smallest perceivable change in magnitude, the difference threshold, is proportional to the magnitude of the stimulus:

$$\Delta I = kI$$

where ΔI is the difference threshold and k is a constant. For example, for simulated automobile seat vibration, Weber's Law has been shown to hold with a difference threshold of about 13% for a variety of road stimuli (Mansfield and Griffin, 2000).

2.2.2 PERCEPTION IN SEATED POSTURES

Most exposure to whole-body vibration occurs in seated postures while people are driving or are passengers in transportation. Therefore, most research into the perception of whole-body vibration has used subjects in seated postures. These studies can use two approaches: either to map the perception threshold with frequency or to map equal-intensity curves with frequency. Usually, studies have been carried out in the laboratory using single-axis sinusoidal signals due to the methodological difficulties in using complex multiaxis stimuli. One common finding is that individuals vary in their perception thresholds, often by a factor of 2:1.

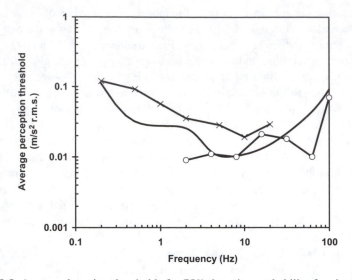

FIGURE 2.2 Average detection thresholds for 75% detection probability for sinusoidal vertical whole-body vibration of seated subjects and perception threshold from BS 6841 (1987) for 50% of alert fit persons. [Mean data from Benson and Dilnot, 1981 (–×–); median data from Parsons and Griffin, 1988 (–○–); threshold values from BS 6841, 1987 (——)].

For vertical vibration, the motion is perceived most easily at about 5 Hz (e.g., Benson and Dilnot, 1981; Parsons and Griffin, 1988; Howarth and Griffin, 1988). Sinusoidal vibration can be perceived at about 0.01 m/s^2 r.m.s. at 5 Hz, equating to a displacement of about 0.01 mm r.m.s. (Figure 2.2). Below 1 Hz, about 0.03 m/s^2 r.m.s. is required for perception; at 100 Hz, about 0.1 m/s^2 r.m.s. is required for perception. Below 0.5 Hz, it is possible to see motion that cannot be perceived using other physiological systems.

For horizontal vibration, the motion is perceived most easily below 2 Hz (e.g., Parsons and Griffin, 1988; Howarth and Griffin, 1988). As for vertical vibration, sinusoidal horizontal motion can be perceived at about 0.01 m/s^2 r.m.s. at the most sensitive frequencies (Figure 2.3). For 1 Hz, 0.01 m/s^2 r.m.s. equates to a displacement of about 0.25 mm r.m.s. Above about 2 Hz, sensitivity decreases to a threshold of about 0.4 m/s^2 r.m.s. at 80 Hz.

BS 6841 (1987) also provides guidance on perception thresholds. This states that 50% of alert and fit persons can just detect a weighted vibration with a peak magnitude of approximately 0.015 m/s^2 (which equates to 0.01 m/s^2 r.m.s. for sinusoidal signals). The standardized threshold value applies to vertical and horizontal vibration.

2.2.3 PERCEPTION IN NONSEATED POSTURES

Despite most whole-body vibration exposures occurring for seated postures, there are many environments where other postures are important. For example, rush-hour travel often involves standing in trains, trams, or buses; travelers sleep in prone, supine, or semisupine postures on ships or aircraft; injured patients are transported

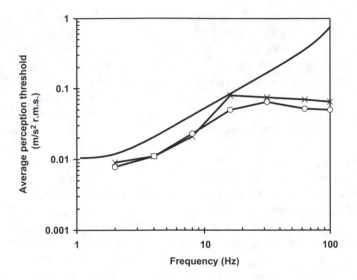

FIGURE 2.3 Median detection thresholds for 75% detection probability for sinusoidal horizontal whole-body vibration of seated subjects and perception thresholds from BS 6841 (1987) for 50% of alert fit persons. [Data from Parsons and Griffin, 1988: fore-and-aft (–×–) and lateral (–○–); threshold values from BS 6841, 1987 (——)].

to and from hospital or are evacuated from military operations in supine postures; and many sleeping areas are exposed to ground-borne vibration from passing heavy trucks or trains.

In standing postures, perception thresholds for whole-body vibration at the floor are generally similar to those for vibration at the seat for seated persons, although there is some evidence for a decreased sensitivity to horizontal vibration when standing (e.g., Parsons and Griffin, 1988). However, due to the similarity of the anatomy of the hand and foot and anecdotal evidence of "vibration white toe" (Pelmear and Wasserman, 1998), it might also be appropriate to assess perception of vibration at the feet using techniques developed for hand-transmitted vibration. (There are some studies of vibration perception thresholds for individual toes in the literature on diabetes, but these have been carried out using methodologies that do not include the support of bodyweight on the feet. There are no known studies of vibration perception thresholds for the feet that have been carried out using techniques similar to those usually employed for whole-body vibration research.)

In recumbent postures, vibration perception thresholds are similar to those for seated postures. It should be noted that vertical vibration of the supporting surface is most easily perceived at about 5 Hz, irrespective of whether that corresponds to the z-axis (seated), x-axis (lying prone or supine), or y-axis (lying on one side) motion for the biomechanical coordinate system (Maeda et al., 1999; Yonekawa et al., 1999).

2.2.4 FREQUENCY WEIGHTINGS

A frequency weighting is a frequency response function that models the response of the body to wave phenomena. Another way of considering frequency weightings is that they are an inversion of an equal response (e.g., perception, pain) curve. In Subsection 2.2.3 it was observed that the seated human is 10 times more sensitive to vibration at 5 Hz than at 100 Hz. Therefore, it could be proposed that measurements of vibration at 100 Hz are reduced by a factor of 10 when compared to measurements of vibration at 5 Hz, in order to maintain subjective sensation parity between the two measures. The concept can be extended such that there is a continuum of weighting factors across the frequency range of possible perception. Frequency weightings are thus applied to time-domain signals to modify them so that the weighted signal represents the human response to the vibration rather than the mechanical characteristics of the vibrating surface. Usually, the procedure applies mathematical digital signal processing techniques, the details of which are beyond the scope of this book (for further reading see Bendat and Piersol, 1986).

There are three limitations of using frequency weightings for whole-body vibration, although these limitations also apply for hand-transmitted vibration. First, frequency weightings are derived from meta-analyses of studies of equal sensation curves. They are therefore representative of a population rather than an individual. As a result, inter- and intra-individual differences cannot be reproduced by using weighting techniques. For example, a householder complaining about residential building vibration might have a very sensitive perception threshold such that he or she can feel the movement, whereas other individuals might not be able to perceive any vibration at all. In this example, vibration measurements might indicate that magnitudes should be below threshold. One is therefore faced with the challenge of whether to design for the population as a whole or for the sensitive extremes in terms of perception (this is analogous to classic anthropometry problems).

A second limitation of human vibration frequency weightings is that they assume linearity, i.e., there is only one weighting that is used for both low- and high-magnitude environments. Howarth and Griffin (1988) show small changes in the shape of the equal-sensation curves at the extremes of their vibration magnitudes and that subjective sensation increased more rapidly than would be expected from Stevens' Law for the lowest vibration magnitudes. Although these differences are slight when compared to interindividual variability, they are an indication of the nonlinearities in human response. Shoenberger (1972) proposes a series of curves for varying vibration magnitudes (analogous to "phon" curves in acoustics), but their use is not widespread.

A third limitation of frequency weightings is that they assume that techniques based on perception (or ratings of equal comfort) can be used for prediction of injury. This assumption is an important consideration because unlike other sensory systems (e.g., visual, auditory), the anatomical locations where injury might occur for whole-body vibration are not the same as the locations where physiological receptors are located. One must question the appropriateness of linking the psychophysiological response of the visual, vestibular, somatic, and auditory systems with the pathological response of spinal tissue.

Although there are problems with the use of frequency weightings, a key consideration is that there is, currently, no alternative method for the assessment of complex vibration with components at multiple frequencies that has proved to be

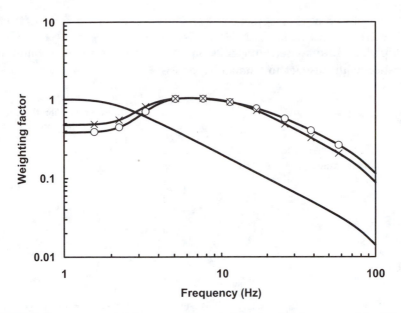

FIGURE 2.4 Frequency weighting curves for whole-body vibration as used by BS 6841 (1987), ISO 2631-1 (1997), and the EU Physical Agents (Vibration) Directive (2002). Vertical vibration: W_b (–o–) and W_k (–×–); horizontal vibration: W_d (——).

better than the use of frequency weightings. For the time being at least, the principle of "if it feels unhealthy then it probably is unhealthy" should be maintained.

For whole-body vibration, the three most commonly used frequency weightings are known as W_b, W_k, and W_d (Figure 2.4). W_b and W_k are both used for vertical vibration only and are each characterized by a peak at 5 Hz. W_k is about 25% higher than W_b below 3 Hz and is lower than W_b at frequencies above 12 Hz. These differences are small when compared to intersubject differences for vibration perception, on which the weightings are based. W_b is used by BS 6841 (1987); W_k is used by ISO 2631 (1997), the EU Physical Agents Directive (2002), and the most recently introduced standards. For most environments, assessments made using W_k will generate higher values for the weighted acceleration than using W_b (e.g., Lewis and Griffin, 1998; Paddan and Griffin, 2002a, b).

It is possible to encounter documents referring to ISO weightings from versions of ISO 2631 prior to 1997. These weightings were different from the current weightings and are often labeled ISO-x, ISO-y, and ISO-z. Other weightings include W_e (for rotational vibration) and W_c (for vibration at the back of the seat). Such techniques are not limited to human response to vibration; weightings have even been developed to assess the welfare of chickens during transportation (e.g., Randall et al., 1997).

2.2.5 VIBRATION PERCEPTION IN BUILDINGS

Buildings might move due to passing heavy transport, seismic activity, or exposure to strong winds. People might respond in a variety of ways, including distraction,

TABLE 2.1
Multiplying Factors Used to Specify Satisfactory Magnitudes of Building Vibration with Respect to Human Response

		Multiplying Factors	
Place	Time	Continuous Vibration	Intermittent Vibration and Impulsive Vibration Excitation with Several Occurrences Per Day
Critical working areas (e.g., some hospital operating theaters, some precision laboratories, etc.)	Day	1	1
	Night	1	1
Residential	Day	2–4	60–90
	Night	1.4	20
Office	Day	4	128
	Night	4	128
Workshops	Day	8	128
	Night	8	128

Source: From British Standards Institution (1984). Evaluation of human exposure to vibration in buildings (1 Hz to 80 Hz). BS 6472. London: British Standards Institution.

fear, annoyance, sleep disturbance or, alternatively, precision activities might be compromised. Occupants are quick to complain of building vibration even if it is only just above perception thresholds. If the movement of the building is sufficient, then slight (or severe) structural damage might occur producing a psychological response beyond the scope of this chapter.

Two technically identical standards, ISO 6897 (1984) and BS 6611 (1985) provide guidance for assessing the motion of tall buildings or fixed offshore structures to low-frequency motion (0.063 to 1 Hz). The standards provide a "satisfactory" magnitude of motion for the peak 10 min of the worst windstorm with a return period of 5 years. For building motion, satisfactory magnitudes are below 0.067 m/s^2 r.m.s. at 0.1 Hz falling to 0.026 m/s^2 r.m.s. at 1 Hz. These values do not consider other cues to the motion such as visual or auditory.

For vibration between 1 and 80 Hz, ISO 2631-2 (1989) and BS 6472 (1984) apply. Their approach is identical but there are slight differences in detailed criteria. A "base curve" is specified that, above 4 Hz, is similar in shape to the "perception threshold" curve in BS 6841 (1987) but is lower between 1 and 4 Hz (i.e., implies a greater sensitivity). Satisfactory magnitudes of building vibration are specified as multipliers of the base curve, depending on the type of activity in the building and whether the vibration is continuous (more than 16 h/d), intermittent, or impulsive (stimuli <2 s). Multipliers for continuous vibration range from 1, for "critical working areas" such as some hospital operating theaters, to 128 for intermittent vibration in workshops (Table 2.1). A multiplier of 2 causes the curve in BS 6472 to replicate the perception threshold curve in BS 6841 above 4 Hz. Therefore, acceptable magnitudes of continuous vibration for nighttime residential and some other uses are below average perception thresholds.

2.3 WHOLE-BODY VIBRATION COMFORT AND DISCOMFORT

Many dictionaries provide a definition of comfort aligned to "an absence of discomfort." In the same dictionaries, discomfort can be defined as "an absence of comfort," leaving us with a circular argument. Physiologically, humans have no comfort receptor despite having a battery of pain receptors (nociceptors). One could therefore define one criterion for comfort as "an absence of signals from nociceptors." However, there is also a range of vibration magnitudes that might be considered uncomfortable but not painful. Perhaps the most appropriate definition of a comfortable stimulus is one where the subjects would not change their activity to reduce its magnitude.

The scope of this section is bound by vibration signals that are above perception thresholds but below magnitudes that might be considered hazardous to health. An important factor for comfort in vehicles is the design of the seat; this is treated separately in Section 2.5.

2.3.1 NOISE, VIBRATION, AND HARSHNESS (NVH)

NVH as a discipline has grown in the automotive industry since the 1980s. It was spawned from an increasing demand from customers for improved vehicle refinement, ride, and quality. NVH engineers use a colorful terminology for noise and vibration including shuffle, idle shake, lugging roughness, and jiggle shake. An objective for NVH is not only to reduce levels of cabin noise and vibration but to tune them so that there is a feeling of smoothness within the car. Therefore, it provides the automotive industry with a crossover point for engineering, ergonomics, and psychology.

One aspiration for NVH engineering is the prediction of comfort levels in the production vehicle without the requirement for extensive prototyping. This has demanded the development of predictive models for perceived ride comfort in addition to the development of seat testing techniques (Section 2.5).

2.3.2 VEHICLE COMFORT

In a "comfortable" vehicle, occupants tend not to actively consider the background vibration despite magnitudes being above perception thresholds. The vibration signals are thus filtered out by the brain. However, if the vehicle passes over a bump in the road, a combination of vibration, auditory, and possibly visual stimuli are perceived as an "event" that is noticed by the occupant. This "event hypothesis" is analogous to the cocktail party effect whereby unimportant sounds and speech are filtered out by cognitive processes.

The perceived discomfort increases as the intensity and number of events increases. Usually, background vibration contributes only a small proportion of the total discomfort if there are events in the vibration, but this depends on the number and nature of the events and the magnitude of the background vibration.

One might assume that changing a vehicle to decrease weighted vibration emissions would lead to an improved vehicle ride. Although this can be true, it depends on the limiting factor for the ride of the original vehicle. In agriculture, for example, tractors are often driven at a speed such that the drivers moderate their discomfort. They will drive at higher speeds on smoother surfaces so that the overall discomfort is maintained. This concept can be further developed by considering the development of fully suspended tractors in the early 1990s that could provide a significant reduction in vibration exposures if speeds were not increased (Scarlett and Stayner, 2001). However, many drivers use the tractors in such a way that speeds are increased, enabling tasks to be completed quicker but also resulting in little or no reduction in the vibration. Quicker completion of tasks should be welcomed as this will reduce the exposure time either of the individual or of the workforce as a whole, so long as the total vibration dose does not increase. This behavior can be termed discomfort homeostasis (analogous to risk homeostasis).

2.3.3 ASSESSING VIBRATION COMFORT

Laboratory studies of whole-body vibration discomfort have established a relationship between the magnitude, duration, frequency content, and waveform of the signal. It is not surprising that a higher magnitude stimulus is more uncomfortable than a lower magnitude stimulus (e.g., Mansfield et al., 2000). For stimuli up to 2 min, longer-duration stimuli are more uncomfortable than shorter-duration stimuli (Griffin and Whitham, 1980; Kjellberg and Wikström, 1985). However, for longer durations, the discomfort increases less rapidly, if at all (Kjellberg and Wikström, 1985).

There are many laboratory investigations of the effect of frequency on comfort in the literature (e.g., Shoenberger and Harris, 1971; Miwa and Yonekawa, 1971; Howarth and Griffin, 1988). Many of these have used semantic rating scales, asking subjects to rate stimuli as, for example, slightly uncomfortable, uncomfortable, and very uncomfortable (e.g., Shoenberger, 1982; Huston et al., 2000). Others have used reference vibration stimuli for comparison to test stimuli (e.g., Griffin et al., 1982). These and other techniques have applied a wide range of psychophysical methods, postures, and shakers, but most show similar trends in their results. Vibration at about 5 Hz is generally reported as producing a greater sensation than lower or higher frequencies. This trend follows the same pattern as that established for vibration perception threshold curves, thus validating the general shape of the frequency weighting curves.

A final factor for vibration comfort is the waveform of the signal. Shocks have been shown to cause more discomfort than other stimulus types of the same frequency-weighted r.m.s. vibration magnitude (Mansfield et al., 2000). This is one reason why assessment techniques that emphasize high acceleration events (such as the vibration dose value) perform better than r.m.s. for assessments of whole-body vibration comfort.

When considering vibration comfort, the interaction between magnitude, duration, frequency content, and stimulus waveform is not trivial and is further confounded by inter- and intra-subject differences and nonvibration factors. For example, the physical environment in a car cabin at 30°C will be considered "too hot," and this response will dominate feelings of discomfort, irrespective of the subtleties of the vibration environment. Various authors have suggested a variety of techniques for assessing comfort using a range of frequency weightings, averaging (or peak detection) techniques, and time dependencies. Some laboratory studies (e.g., Mansfield et al., 2000; Ruffell and Griffin, 2001) and field studies (e.g., Wikström et al., 1991) have identified that dose measurements based on the fourth power of the frequency-weighted acceleration are superior, or at least not inferior, to using second power measures (i.e., r.m.s.) for the prediction of discomfort. Therefore, application of the vibration dose value (VDV) is recommended above r.m.s. or peak methods for assessment of vehicle comfort.

2.3.4 ACTIVITY INTERFERENCE

The demands of a time-pressured culture in combination with advances in information and communications technology has led to an increase in the range of tasks performed while traveling. Reading, writing, and eating can be affected by whole-body vibration. The extent of the disturbance depends on the nature of the vibration (e.g., frequency, magnitude, direction, waveform; see Figure 2.5). The problem of writing has, to some extent, been eliminated by the developments in mobile computing allowing for use of word processors while traveling. Although one might also expect typing to be affected by vibration, there are no known studies investigating the extent of its effects. Increased functionality for mobile telephones means that small keys are now used not only for dialing but for accessing information and typing short messages, introducing a new arena of activity interference from whole-body vibration. It is impossible to accurately predict the longevity and development of such technologies, and so experimental laboratory data–based improvements in usability might not be achievable due to the insatiable appetite for new products (at least by the suppliers) driving rapid technological changes.

2.4 HEALTH EFFECTS OF WHOLE-BODY VIBRATION

The most commonly reported health effect of whole-body vibration is back pain. Indeed, back pain is the focus for the rest of this section. Despite this, other types of health effects have been observed. These include sciatica, digestive disorders, genitourinary problems, and hearing damage (Griffin, 1990). Portuguese researchers have reported links between low frequency noise and whole-body vibration with a variety of disorders, collectively termed vibroacoustic disease (e.g., Castelo Branco and Rodriguez, 1999; Castelo Branco, 1999). These include thickening of cardiac structures and neurological and vascular disorders. Despite the name of the condition, the greatest contributor to risk is considered to be large-pressure-amplitude ($\geq$90 dB SPL), low-frequency ($\leq$500 Hz) noise and not whole-body vibration as described in this book. One might argue that high-intensity, short-duration impacts that can cause trauma of the internal organs is a form of whole-body vibration. Such topics fall within the remit of crash research and are beyond the scope of mainstream whole-body vibration research. At the other extreme, low-intensity, long-duration vibration might cause annoyance leading to elevated blood pressure and other stress-related symptoms.

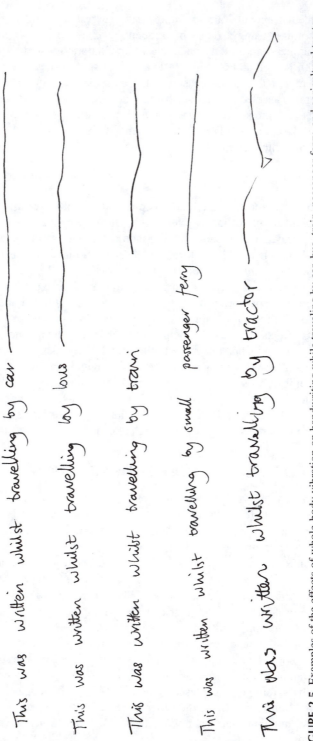

FIGURE 2.5 Examples of the effects of whole-body vibration on handwriting while traveling by car, bus, train, passenger ferry, and agricultural tractor.

A problem with many of the symptoms previously described is that they might be caused by a variety of factors. Consider two journeys, one in a new luxury car and one in an old budget car. It is likely that the occupants will feel more fatigued after the drive in the old car. If the vibration magnitudes in the cars were measured, then it might also be observed that the exposure was greater in the old one. Does this mean that the fatigue was caused by the vibration? Perhaps it was. However, it might also have been caused by the design and condition of the seat, the quality of the posture, or one of a host of other differences between the cars. As issues of whole-body vibration and health are considered, it is important to remember that vibration is one of many risk factors that dictate the need for a holistic approach.

2.4.1 EPIDEMIOLOGICAL STUDIES

The problem of linking a specific pathogen to a specific disease is not exclusive to whole-body vibration. Epidemiologists have developed techniques that are helpful in establishing links. Hill's criteria (see Table 2.2; Bradford-Hill, 1966) provide a nine-part epidemiological framework for establishing a link. It is important to note that the only essential criterion is the temporal relationship; all other criteria help to support the relationship, but none determine a link irrefutably. For considering links between whole-body vibration exposure and back pain, the first five criteria are the most useful (i.e., strength of association, consistency of association, plausibility of relationship, temporal relationship, biological gradient). The other criteria are sometimes difficult to test. It is, for example, unethical to attempt to deliberately

---

TABLE 2.2
Hill's Criteria for Pathogen-Disease Causality

Criteria	Example of Criteria
1. Strength of association	There is a strong link between the pathogen and the disease, with the risk of exhibiting symptoms substantially increased with exposure to the pathogen (e.g., supporting odds ratios)
2. Consistency of association	There is a consistent body of evidence with repeated studies linking the pathogen and disease
3. Plausibility of relationship	The association between the pathogen and disease is plausible, considering the biomechanics of the system
4. Temporal relationship	The disease occurs after exposure to the pathogen and not before
5. Biological gradient	An increase in the severity of the disease is observed with an increase in the exposure to the pathogen (e.g., a dose–response relationship)
6. Coherence	The link between disease and pathogen is in agreement with the existing state of knowledge
7. Experimental evidence	Initiation of the disease can be demonstrated through laboratory simulation or administration of the pathogen
8. Analogy	There is knowledge of similar processes within the biological sciences
9. Specificity	A unique link between the pathogen and the disease can be identified, i.e., a lack of other pathogens

cause injury through vibration exposure (as required for experimental evidence criteria). Similarly, other common pathogens are known for back pain (specificity criteria).

Epidemiological data for specific exposure types have been repeatedly reported since the 1950s. Griffin (1990) summarizes over 130 studies of whole-body vibration and health between 1949 and 1988. It is helpful, therefore, that epidemiological studies of whole-body vibration and health are reviewed periodically (Hulshof and van Zanten, 1987; Griffin, 1990; Kjellberg et al., 1994; Bovenzi and Hulshof, 1998; Lings and Lebouef Yde, 2000; Stayner, 2001). There is general agreement between these reviews that there is evidence for a link between vibration exposure and low back pain. Bovenzi and Hulshof (1998) state:

> Occupational exposure to whole-body vibration is associated with an increased risk of low back pain, sciatic pain, and degenerative changes in the spinal system including lumbar inter-vertebral disc disorders.

The authors combined data from two previous studies that indicated that, for tractor drivers, an increased prevalence odds ratio was observed for increased vibration exposures, suggesting a link between back pain and vibration dose (Figure 2.6).

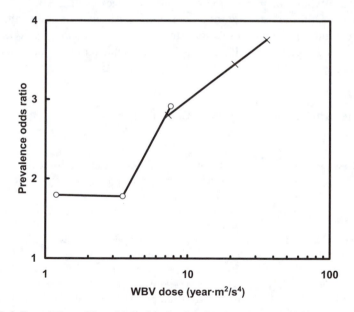

FIGURE 2.6 Prevalence odds ratio for low back pain among tractor drivers as a function of lifetime cumulative whole-body vibration (WBV) dose estimated as $\Sigma a_i^2 t_i$, where a_i is the root sum of squares of the frequency-weighted r.m.s. acceleration of tractor i and t_i is the number of full-time working years of driver on tractor i (year m²/s⁴). Data from Boshuizen et al. 1990 (–o–) and Bovenzi and Betta (1994) (–×–). [Adapted from Bovenzi, M. and Hulshof, C.T.J. (1998). An updated review of epidemiologic studies on the relationship between exposure to whole-body vibration and low back pain. *Journal of Sound and Vibration*, 215(4), 595–612.]

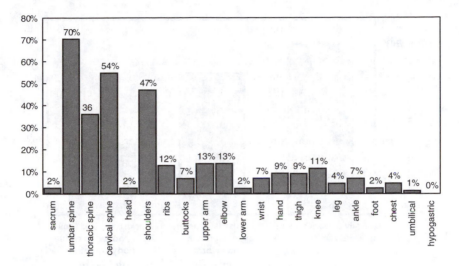

FIGURE 2.7 Prevalence of pain, aching, or discomfort following rallying for 90 drivers and co-drivers. [Adapted from Mansfield, N.J. and Marshall, J.M. (2001). Symptoms of musculoskeletal disorders in stage rally drivers and co-drivers. *British Journal of Sports Medicine*, 35, 314–320.]

Mansfield and Marshall (2001) performed a questionnaire study to investigate the prevalence of musculoskeletal troubles experienced by rally drivers and co-drivers (Figure 2.7). These data were compared with a meta-analysis of reported prevalence of troubles in those occupations traditionally associated with whole-body vibration exposure (e.g., forklift truck drivers, tractor drivers) and also with controls who were not employed in such jobs (e.g., car drivers, office workers). This study had the advantage of using subjects who were exposed to high magnitudes of vibration and mechanical shock but for short periods of time (i.e., rally drivers and co-drivers), rather than those exposed to low or moderate magnitudes of vibration but for extended periods of time (i.e., the two groups used for comparison). Furthermore, pain was specifically related to the rally vibration exposure. A clear increase in prevalence of pain was shown for the rally participants when compared to the other two groups (Figure 2.8).

In the U.K., it is estimated that 9 million workers are exposed to whole-body vibration each week (Palmer et al., 1999). Of these, 383,000 are estimated to be exposed to magnitudes of vibration that exceed the 15-VDV guidance level from BS 6841 (1987). Occupations with the most people exposed at this level are farming, drivers of road goods vehicles, and forklift and industrial truck drivers.

Most of the epidemiological studies reported in this section have used questionnaire-based evidence to identify back trouble. However, if injury occurs, then tissue damage would be expected, thus resulting in a pathological change.

2.4.2 PATHOLOGICAL STUDIES

Despite back pain being reported across a variety of occupations with vibration exposure, it has proved difficult to diagnose any specific disorders of those employed

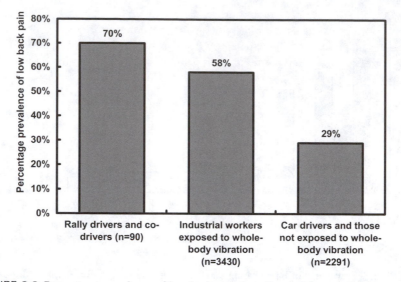

FIGURE 2.8 Percentage prevalence of low back pain for rally participants, those exposed to industrial whole-body vibration, and those not exposed to whole-body vibration (data from Mansfield and Marshall, 2001).

in such industries, or even for those reporting pain. Drerup et al. (1999) used a cohort of 20 heavy earthmoving machine operators who had clinically verified back complaints and had applied for early retirement. All operators had more than 15 years' exposure of whole-body vibration. The authors applied a battery of tests (e.g., stadiometry, MRI imaging, CT scans) in an attempt to find pathological changes to the spine. No significant differences were observed between the cohort with long-term vibration exposure and a group of nonexposed controls. Indeed, the authors concluded that there was:

> No evidence that lumbar discs of subjects exposed to long-term whole-body vibration differ on average from those of nonexposed subjects.

Similarly, Videman et al. (2000) used MRI imaging to investigate the spinal pathologies of 18 rally drivers, of which 89% reported low back pain. As per Drerup et al.'s study, no differences were observed between the rally drivers and controls:

> … Even extreme vehicular vibration as associated with rally driving does not appear to have significant effects on disc degeneration …

A key factor with these two studies is that virtually all of the participants reported back pain. Therefore, from an individual level, the problem of pain exists, despite the current lack of technology to identify the specific pathology that causes the pain.

2.4.3 PROTECTION

The previous sections have highlighted the association between whole-body vibration and risk of musculoskeletal troubles. Occupational health good practice dictates that agents inducing a risk should be reduced. For vibration protection, there are only two strategies available: either to reduce the duration of the exposure or to reduce the magnitude of the exposure. Changing working practices entirely such that machines are remote controlled and operator vibration is eliminated is perhaps the ideal (but rarely feasible) solution, so long as other ergonomic risk factors are not increased as a result.

The first method of protection should be to implement engineering solutions to the problem. To reduce vibration at source, it is important to understand where the major contributor to the vibration originates.

For some environments (e.g., forklift trucks) a major contributor to the vibration is the roughness, or tidiness, of the floor or road. In these situations, maintenance of floors and keeping debris from the vehicle routes would improve the mechanical (vibration) environment for the vehicle. There might also be other spinoff benefits such as elimination of trip hazards for pedestrians. Some areas of a workplace could be prohibited for certain machines, either to preserve quality floors or to avoid rough ones. For many vehicles, a lower vibration dose is received by the operator if the speed of the vehicle is limited.

Occasionally, the engine of the vehicle can be the main contributor to the vibration exposure. Considering that the body is most sensitive to whole-body vibration at frequencies less than 20 Hz, it is best for single-cylinder engines to run at speeds greater than 1000 rpm. The more cylinders that exist in the engine, the higher the dominant frequencies of vibration. Therefore, it is often better to use multicylinder engines. Engines should be mounted on resilient mountings to isolate the engine vibration from the chassis of the vehicle and for engines to be balanced, where possible.

Tools used by the machine can be a source of vibration. Sometimes vibration is an essential part of the process (e.g., ride-on rail grinders, hydraulic breaker attachments, driving or piling machines, tamping machines). In these cases the frequency of vibration should be selected to be effective for the task but out of the sensitive range of the operator. Other tools (e.g., backhoe loader excavators) are hydraulically driven and the mechanisms should be designed such that a smooth operation is achieved. The recent move from direct linkage to servo control for some types of machine provides opportunities for development of idealized mechanisms for vibration reduction. Finally, the correct tool should be used for the task (e.g., excavator buckets should not be used for piling).

If the source of the vibration cannot be adequately eliminated, then the transmission of vibration to the operator should be minimized. This can be achieved by isolation of the source (e.g., axle suspension, engine mounting, tool isolation) or isolation of the operator (e.g., suspended cabs, suspension seats).

If exposures cannot be reduced by engineering means, then reducing the number of operations per exposed person might be the only option for reducing vibration exposure dose. This could be achieved by reducing demands on the operator, allowing for a slower completion rate, or by job (or task) rotation. Job rotation is not always effective at reducing risk. For example, some machines might expose the operator to most vibration while traveling from a store to a work site (e.g., a road roller) and so increasing the number of such journeys, albeit with different operators, might not be an effective means of reducing vibration exposure.

Through innovation it is possible to reduce vibration exposure while improving performance (e.g., agricultural machines with suspended axles). Vehicle manufacturers and component suppliers are often aware of critical parts of their designs that could improve the vibration efficiency, at a cost. If customers along the supply chain consider vibration performance as a purchasing criterion, then, to maintain a competitive advantage, manufacturers may improve the technologies employed for vibration isolation. As a result, operators should be exposed to lower magnitudes of whole-body vibration.

2.4.4 THE "HOLISTIC" APPROACH

The cause of an individual's back pain is often elusive. Most researchers agree that vibration is a risk factor for back pain, but there are other factors that are important. When making an assessment of an environment, it is therefore essential that the entire task be considered. Other risk factors like poor posture, prolonged sitting, heavy lifting, and working in the cold are often associated with environments where whole-body vibration is also present. It is not uncommon to observe drivers jumping down from cabs of trucks or agricultural machines; this action might be more of a risk factor than the exposure to the vibration. Consider helicopter pilots as an example; there are few researchers who disassociate the occupation with back pain, but if vibration were the only cause of pain, then one would expect similar prevalence of pain for pilots and copilots. Bridger et al. (2002a,b) report a significantly lower prevalence of back pain for copilots (24%) than for the pilots (up to 72%, depending on the task) and conclude that the poor posture of the pilot is likely to be another important risk factor for the cause of the pain. Similarly, Magnusson and Pope (1998) comment that vibration is only one of many pathogens and recommend that core ergonomic factors be considered with any vibration assessment.

Stayner (2001) highlights that although it is possible to associate back pain with an occupation, it is far more difficult to identify which aspect of the occupation is the cause of the pain. He makes the observation that:

> It is a common feature of many of the research reports and reviews that they refer to occupational groups as "exposed" groups, using "WBV exposure" as a synonym for working on a machine, and thereby implying that they are concerned only with vibration exposure. ... This seemingly innocuous mistake carries with it the assumption that the researchers are not trying to discover the cause of the disease or effect, but that they have already decided what it is and are intent only on finding evidence in support of their decision. This is perhaps an inappropriate basis on which to conduct scientific research.

This is a fair criticism of much of the previous research and forms the basis for some of the challenges facing the next generation of whole-body vibration research.

So, does vibration cause back pain? If Hill's criteria are considered (Table 2.2), then one can confidently conclude that the evidence points toward an association based on most criteria. The problem criterion is specificity, but the existence of other risk factors does not preclude vibration from being a pathogen.

Whether vibration or posture or lifting or any other factor caused the back pain for any individual is almost impossible to categorically ascertain, but a holistic assessment (i.e., considering the "whole" environment) can be helpful in estimating (and reducing) the relative risk factors.

2.5 SEATING DYNAMICS

Vehicle seats are required to perform a variety of tasks. For example, a seat should position the driver such that they can operate the vehicle safely and effectively; it should have strength enough to protect the occupant in the event of an accident; it should be adjustable such that the driver can position himself or herself in a comfortable posture; it should have good static comfort properties; it should have good dynamic comfort properties. For some seats, thermal properties are important; others might incorporate secondary safety features such as airbags. The seat's aesthetic properties are essential for many applications. Of course, many of the requirements are also applicable to vehicle passengers in addition to the drivers who may, in some cases, be the ultimate customers (e.g., luxury cars, public transport). It is the job of the seat design team to provide the optimal compromise of these factors in a cost-effective way.

2.5.1 OPTIMIZING VEHICLE SEATING COMFORT

Vehicle seating comfort has two aspects: static parameters and dynamic parameters. Some of the static parameters investigated in the literature include driver posture (e.g., Reynolds, 1993; Porter and Gyi, 1998), contact pressures (e.g., Shen and Parsons, 1997; Gyi and Porter, 1999), and thermal properties (e.g., Parsons, 2003).

Vehicle ergonomists are often asked, "What is the most comfortable posture?" by designers. One answer to this question is: "The next one!" It is natural to continually change postures to use and rest alternative muscle groups; therefore, a comfortable seat will allow the occupant some movement and a range of possible body positions. This ideal is difficult to achieve, especially for car seats, as they are also designed to protect the occupant in the event of a crash whereby seat restraints restrict movement. Designers can use generic guidance published by bodies such as the Society of Automotive Engineers (SAE), although such guidance is based on anthropometric accommodation rather than long-term comfort (these guidelines are published in the annually updated SAE handbook). Contact pressures can be mapped and provide objective measures of localized strain related to seat-foam stiffness. For short-term evaluations, the pressures measured beneath the ischial tuberosities correlate to discomfort (Ebe and Griffin, 2001) although the agreement has not been reproduced for long-term sitting (Gyi and Porter, 1999).

Ebe's model of seat discomfort (Ebe and Griffin, 2000a,b) includes a consideration of both static and dynamic factors. The model is formed by addition of the discomfort caused by the static factors to the discomfort caused by the dynamic factors (Figure 2.9a). For a car that is not moving, there is zero vibration (this corresponds to assessments of comfort made in the showroom without test driving a new car and when first impressions of seating comfort are made). When there is

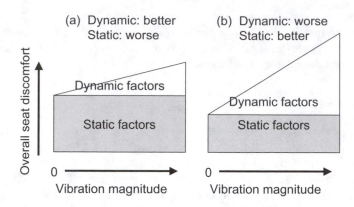

FIGURE 2.9 Ebe's model of overall seat discomfort showing the changing relative importance of static and dynamic parameters depending on the vibration magnitude. [Adapted from Ebe and Griffin (2000a). Qualitative models of seat discomfort including static and dynamic factors. *Ergonomics,* 43(6), 771–790.]

no vibration, comfort is dictated by the static properties. As the vibration magnitude increases, the relative importance of the dynamic characteristics of the seat increases. Dynamic discomfort increases more rapidly for seats with poorer dynamic factors. In Figure 2.9, seat (a) causes more static discomfort than seat (b). However, the rate of increase of dynamic discomfort is greater for seat (b) than for seat (a). Therefore, seat (a) will be better for environments with large magnitudes of vibration but seat (b) will be better of environments with low magnitudes of vibration. It should be noted that the total discomfort (static plus dynamic) increases with the vibration magnitude for both seats.

When approaching the design of a vehicle seat, the vibration in the environment for which the vehicle is designed should be considered. This means that, for example, the dynamic performance of a seat destined for a road car with a refined suspension does not have the same relative importance as for a seat destined to be fitted into a four-wheel drive offroad vehicle in terms of overall seat comfort.

Prolonged sitting in one posture will result in a gradual increase in static discomfort irrespective of whether vibration is present (Messenger, 1992). Additionally, the dynamic properties of the seat cushion might change as the temperature of the seat changes to body temperature or the humidity within the foam structure changes due to absorption of sweat. As Ebe's model does not include a time axis, it will not provide a full picture of the changes in discomfort over a long period. However, the model could be extended into a third dimension to include temporal factors (Figure 2.10). As seat occupant discomfort increases due to fatigue with prolonged sitting, the overall discomfort increases accordingly.

Many cars are purchased on the basis of comfort in the showroom. According to these overall models, this is clearly unsatisfactory as it will neglect both the dynamic and temporal aspects of seating comfort. Purchasers should be encouraged to reserve their evaluation of comfort until the ultimate end user has had the opportunity to test-drive the vehicle for an extended period of time, ideally extending to many hours. This is particularly important for high-mileage business drivers.

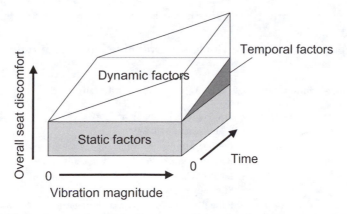

FIGURE 2.10 An improved model of car seat discomfort including static, dynamic, and temporal factors.

Once a car has been delivered, it is difficult and expensive to change the dynamic performance of the vehicle by altering the seats or suspension. Nevertheless, one important aspect of the optimization of vehicle seat comfort remains in the control of the end user: the ability to ensure that the seat is adjusted appropriately. The U.K. Chartered Society of Physiotherapy provides step-by-step guidance on adjusting a seat to its optimal position, although it notes that many cars will not allow for an ideal position for the driver, possibly resulting in a compromised "coping" posture (Table 2.3; The Chartered Society of Physiotherapy, 2001).

2.5.2 CONVENTIONAL SEATS

A conventional seat is one that does not have its own independent suspension mechanism. Conventional seats are usually constructed of a steel frame, polyurethane foam cushions, and a fabric covering. Often, some of the geometry of the seat will be adjustable (e.g., seat height, backrest angle, fore–aft adjustment), and sometimes seats might include heating elements or even a massaging action.

It is common on older trains to observe that the passengers appear to bounce in their seats while the carriage does not appear to move a great deal. In newer trains, the design and dynamic properties of the seats are different, and this phenomenon is less obvious. The reason for the difference in response is due to the different dynamic characteristics or different "transmissibility" of the seats.

Transmissibility is defined as the ratio of the vibration on the seat surface to the vibration at the seat base (usually the floor of the vehicle) as a function of frequency:

$$T(f) = \frac{a_{seat}(f)}{a_{floor}(f)}$$

where $T(f)$ is the transmissibility, $a_{seat}(f)$ is the acceleration on the seat, and $a_{floor}(f)$ is the acceleration at the base of the seat at frequency f. If there is the same magnitude of acceleration at the floor and on the seat surface, then the transmissibility is unity,

TABLE 2.3
Driving Position and Posture Guide for Optimization of Seat Adjustments in a Car

Step	Description	Detail
1	Set seat to initial setup position	• Steering wheel fully up and forward • Seat height at its lowest • Cushion tilted so that front edge in lowest position • Backrest approximately 30° reclined from vertical • Lumbar adjustment backed off • Seat fully rearwards
2	Set seat height	• Raise the seat as high as is comfortable to improve your vision of the road • Check you have adequate clearance from the roof • Ensure you have maximum vision of the road
3	Set fore–aft adjustment	• Move the seat forwards until you can easily fully depress the clutch pedal and accelerator pedal • Adjust seat height as necessary to give good pedal control
4	Set cushion tilt angle	• Adjust cushion tilt angle so that the thighs are supported along the length of the cushion • Avoid pressure behind the knees
5	Set backrest angle	• Adjust backrest so it provides continuous support along the length of the back and is in contact up to shoulder height • Avoid reclining the seat too far as this will cause excessive forward bending of the head and neck, and you may feel yourself sliding forwards on the cushion
6	Set lumbar support	• Adjust lumbar support to ensure even pressure along the length of the backrest • Ensure lumber support "fits" your back, is comfortable with no pressure points or gaps
7	Set steering wheel	• Adjust the steering wheel rearwards and downwards for easy reach • Check for clearance for thighs and knees when using pedals • Ensure display panel is in full view and not obstructed
8	Set head restraint	• Adjust the head restraint to ensure the risk of injury is reduced in the event of a car accident
9	Fine tune	• Repeat stages 2 to 8 and fine tune as necessary

Source: Adapted from The Chartered Society of Physiotherapy (2001). *Take the Pain Out of Driving.* London: The Chartered Society of Physiotherapy.

as would be experienced if the seat was rigid. If the seat is providing isolation at some frequencies, then the transmissibility will be less than unity; all compliant seats also amplify vibration at some frequencies where the transmissibility is greater than unity. As transmissibility is a function of vibration frequency, it is usually presented in graphical form enabling the user to identify where the seat has the greatest response (the resonance frequency) and where the seat provides effective

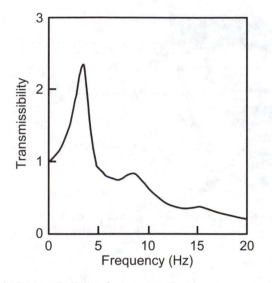

FIGURE 2.11 Typical transmissibility for a conventional seat measured in the laboratory using random vertical vibration (data from Mansfield, 1993).

isolation (Figure 2.11). Almost all conventional seats have a resonance frequency at about 4 Hz and provide isolation above about 6 Hz when loaded with a human adult.

Seat transmissibility alone is not sufficient for providing a complete picture of the effectiveness of the isolation. If there is negligible vibration energy at any frequency, then the transmissibility of the seat becomes unimportant. Similarly, if the vibration occurs at a high frequency where physiological perception mechanisms are not sensitive, again the transmissibility of the seat is unimportant. One convenient way of including all of these factors into a single index is to use the seat effective amplitude transmissibility or "SEAT value" (Griffin, 1990) (it is unfortunate that the term is pronounced "see-at" as there is no logical reason for this potentially confusing convention). The SEAT value inherently includes the three important factors for seat dynamic performance: vibration spectrum, transmissibility, and human response frequency weighting. The SEAT value is defined as (Figure 2.12):

$$\text{SEAT\%} = 100 \times \frac{\text{ride (dis)comfort on seat}}{\text{ride (dis)comfort on floor}}$$

Ride comfort can be quantified using either the r.m.s. or the VDV and therefore the expression becomes either:

$$\text{SEAT\%} = 100 \times \frac{r.m.s._{seat}}{r.m.s._{floor}}$$

or

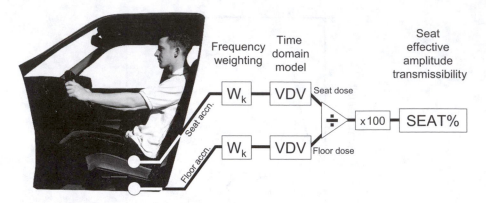

FIGURE 2.12 Graphical representation of the process for calculating the seat effective amplitude transmissibility (SEAT value) for a car seat using the W_k frequency weighting and the vibration dose value (VDV) method.

$$SEAT\% = 100 \times \frac{VDV_{seat}}{VDV_{floor}}$$

respectively. A SEAT value of 100% indicates that the dynamic properties of the seat have not improved or reduced the ride comfort on the seat; a SEAT value of greater than 100% indicates that the ride is worse in the seat than on the floor, and a SEAT value of less than 100% indicates that the dynamic properties of the seat have been effective in reducing the vibration.

Consider two seats, each exposed to two different vibration stimuli (Figure 2.13). Seat A has a resonance frequency of about 3 Hz and seat B has a resonance frequency of about 5 Hz. Vibration stimulus 1 is dominated by vibration between 4 and 8 Hz, and vibration stimulus 2 is dominated by vibration between 0 and 4 Hz. For stimulus 1, seat A is clearly superior to seat B as the spectrum on the surface of the seat is lower at frequencies where there is substantial vibration energy. In this case, the SEAT value for seat A is 80% and the SEAT value for seat B is 140%. For stimulus 2, seat A is clearly inferior to seat B as the spectrum on the surface of the seat is greater at frequencies where there is substantial vibration energy. In this case, the SEAT value for seat A is 130% and the SEAT value for seat B is 110%. This example demonstrates the importance of both the seat transmissibility and the vibration spectrum on the SEAT value and the seat comfort.

What constitutes a "good" SEAT value depends on the vehicle type (Griffin, 1998c). For example, vibration in cars usually has substantial components at about 10 Hz, which can be easily isolated by conventional seats. Car seats therefore often have SEAT values in the range of 60 to 80%. In railway carriages, vibration is dominated by low-frequency motion where conventional seats are unable to provide isolation. Therefore, even a good railway seat might have a SEAT value of greater than 100%.

The passengers riding in the train considered at the start of this section were subjected to a combination of low-frequency vibration exciting a seat that amplifies

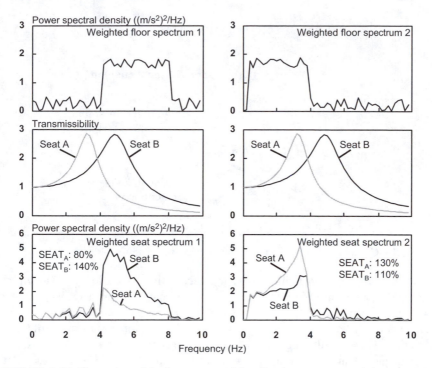

FIGURE 2.13 Illustration of the importance of the vibration spectrum and the seat transmissibility on the seat effective amplitude transmissibility (SEAT value) and on the selection of the seat with superior dynamic performance for the vibration environment.

such motion. Modern train seats are far simpler in construction, consisting of a simple foam cushion, but this is far more effective dynamically than the previous "spring cushion" type seats, therefore producing a lower vibration exposure on the seat surface and, as a result, a lower SEAT value (Corbridge et al., 1989).

2.5.3 SUSPENSION SEATS

A suspension seat consists of an independent suspension mechanism in additional to a conventional foam cushion (Figure 2.14). The suspension mechanism is constructed from springs and a damper that can be mounted beneath the seat cushion or sometimes behind the backrest. Component selection is based on the type of vehicle the seat is designed for (e.g., compressed air springs are often used in trucks, but require an air line). The suspension mechanism increases the size and weight (and cost) of the seat and therefore is only suitable for applications where these can be accommodated. For example, suspension seats are common for large off-road vehicles, where the seat is a small percentage of the total machine mass (and cost), but are impractical for mounting in road cars.

Suspension seats are designed to isolate the occupant from vibration and impacts by altering the seat transmissibility. The suspension mechanism is tuned such that its resonance frequency is about 2 Hz, thereby providing isolation to vibration above

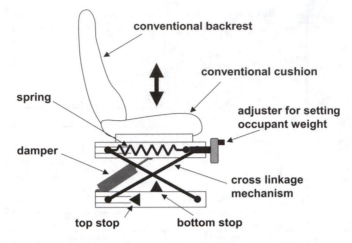

FIGURE 2.14 Component parts of a typical suspension seat.

about 3 Hz. Corbridge (1981) illustrates the effect of the suspension mechanism by measuring the seat transmissibility for a suspension seat with the suspension active and with the suspension inactive (Figure 2.15). For the suspension inactive condition, the seat effectively became a conventional seat, as the cushion was of standard foam construction. These results show that the suspension seat isolates vibration from a lower frequency than the conventional seat. In the example, the seats had a similar performance below 2 Hz, but the suspension mechanism causes the seat to perform better at higher frequencies. At about 5 Hz, there is a 2:1 difference between the transmissibilities for the seat with the suspension active and inactive. It is clear, then, that for a vehicle with vibration-dominated motion in the critical frequency range of 3 to 10 Hz, a suspension seat would reduce the exposure of the occupant.

A problem arises when the mechanisms in suspension seats reach the end of their travel (Wu and Griffin, 1998). If the vehicle traverses a large obstacle, then the limits on the travel might be exceeded causing an impact as the suspension mechanism hits the limit of its compression or extension (Figure 2.16). Considering that impacts are a severe form of vibration, it is possible that the health risk from the impact would be greater than the health risk from the vibration that has been isolated by the seat. To minimize the acceleration during impact, rubber end-stop buffers are mounted at the extremes of the travel for the seat. Hence, the impacts are termed "end-stop impacts." Herein lies a paradox: inserting end-stop buffers into the suspension mechanism reduces the travel of the seat, and therefore increases the incidence of end-stop impacts (albeit of a lower severity).

Wu and Griffin (1996) provide a five-stage model of the performance of a suspension seat with respect to input vibration magnitude (Figure 2.17). At very low magnitudes (stage 1), the acceleration is too low to overcome the friction in the system (sometimes known as "lockup"), and so the suspension provides no isolation [the SEAT value for the suspension part of the seat calculated using the VDV method ($SEAT_{susp}$) is 100%]. At slightly higher magnitudes (stage 2), the suspension starts to move, but nonlinearities, stick-slip characteristics, and "play" in the mechanism

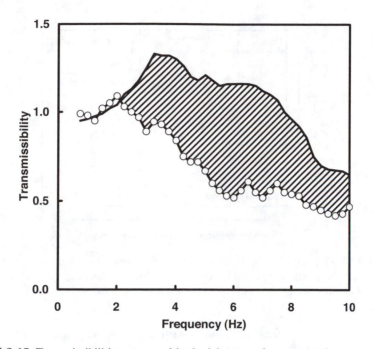

FIGURE 2.15 Transmissibilities measured in the laboratory for a suspension seat with suspension active (–○–) and suspension inactive (——). Shaded area indicates zone where suspension mechanism provides isolation from the vibration (data from Corbridge, 1981).

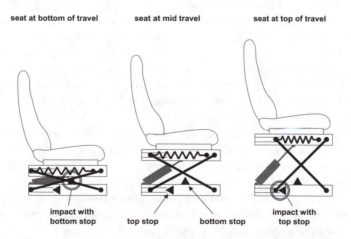

FIGURE 2.16 Possible positions of a suspension seat showing the seat at mid-travel and in positions causing end-stop impacts at the top and bottom stops.

can increase the SEAT$_{susp}$ for a critical range of magnitudes. However, above a threshold, the suspension begins to provide some isolation. At higher magnitudes still (stage 3) the nonlinearities and friction become negligible and the seat performs ideally, providing isolation from the vibration and a SEAT$_{susp}$ of less than 100%. In

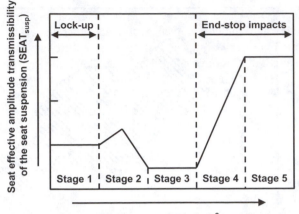

FIGURE 2.17 Five stages of the SEAT value for the suspension mechanism of a suspension seat. [Adapted from Wu, X. and Griffin, M.J. (1996). Towards the standardization of a testing method for the end-stop impacts of suspension seats. *Journal of Sound and Vibration,* 192(1), 307–319.]

stage 3, the SEAT$_{susp}$ does not change with vibration magnitude. The range of vibration magnitudes encompassed by stage 3 performance should be maximized and should include all environments for which the seat is designed. If the magnitudes of vibration are sufficient, then end-stop impacts start to occur and stage 4 is reached where the SEAT$_{susp}$ rapidly increases with increased vibration magnitude. The design of the end-stop buffers should be optimized to reduce the gradient of the increase in SEAT$_{susp}$ in stage 4. The final stage (stage 5) occurs when the end-stop buffers fail to provide any protection from the impacts and the SEAT$_{susp}$ is constant with vibration magnitude. It should be noted that this model is only valid for frequencies of vibration where the suspension mechanism provides isolation; an alternative form of the model is provided by Wu and Griffin (1996) for vibration that coincides with the resonance frequency of the suspension.

There has been some experimentation with using active components in suspension seats (McManus et al., 2002). Such systems constantly monitor the movement of the seat and change the component properties in real time. For example, the damper might utilize magnetorheological fluids to enable the damping to be electronically controlled and increased as the seat travels towards its extreme positions. This control strategy provides more resistance to movement at the extremes of travel such that end-stop impacts can be avoided.

2.5.4 SEAT TESTING

Subsection 2.5.2 showed that the ideal seat in any situation is a function of the seat static properties, the dynamic properties, the vibration that is to be controlled, and the human perception characteristics. Unfortunately, the dynamic response of the seat is influenced by the dynamic properties of the occupant and so cannot be replaced with an inert load. Furthermore, seats are nonlinear such that their dynamic response is a function of the vibration characteristics. Therefore, the transmissibility of a seat measured with one occupant will be slightly different when tested with a different occupant (e.g., Wei and Griffin, 1998). Similarly, the transmissibility of the seat measured while driving on a smooth surface will be slightly different when tested on a rough surface (Fairley, 1983).

One way of removing intersubject variability is to replace the occupant with a dummy that models the dynamic response of the human (see also Subsection 2.6.5.2). A variety of anthropodynamic dummies have been developed for this purpose (e.g., Mansfield and Griffin, 1996; Cullman and Wölfel, 2001; Lewis and Griffin, 2002). Most of these dummies are incapable of reflecting either the differences between occupants or the nonlinear response of the occupant. However, they can be used to provide a repeatable and representative load for seat testing in the lab where man-rated shakers (i.e., vibration simulators that are safe for human experimentation) are unavailable. There are currently no dummies that represent the response of the human in horizontal axes, although there is no technical reason why these might not be developed in the future.

2.6 BIOMECHANICAL RESPONSES TO WHOLE-BODY VIBRATION

It is possible to investigate the response of humans in a dynamic environment by their mechanical responses. The two principal approaches are to make assessments at the "driving point" (i.e., the site of the human contact with the loading force) or remote from the driving point (Figure 2.18). At the driving point, mechanical impedance methods use measures of force and acceleration (or velocity) to determine the response of the body as a whole. Measurements of acceleration remote from the driving point are usually used in combination with simultaneous measures

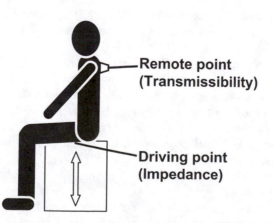

Remote point (Transmissibility)

Driving point (Impedance)

FIGURE 2.18 Illustration of conceptual differences between transmissibility and impedance for biomechanical vibration research.

at the driving point to calculate how vibration is transmitted through the body (transmissibility).

(Occasionally, and especially in older literature, this subject is referred to as "biodynamics." The term is not used in this book to avoid confusion with the increasingly common meaning of the word, which refers to sustainable and organic farming.)

2.6.1 Transmission of Vibration Through the Body

If a person is exposed to a sinusoidal signal that gradually increases in frequency (swept sine), then different parts of the body will resonate in turn. Many body parts will resonate at about 5 Hz (e.g., the head and abdomen), and others at higher frequencies (e.g., the eyeball resonates at about 20 Hz). The extent of the movement at any point on the body is related to the magnitude of the input vibration at the seat or floor and the transmissibility at the driving frequency.

Transmissibility is defined as the ratio of the vibration measured between two points (usually the driving point and a remote location). For seats, transmissibility has previously been defined as (Subsection 2.5.2):

$$T(f) = \frac{a_{seat}(f)}{a_{floor}(f)}$$

If transmission of vibration from the seat surface to the spine is considered, then the equation changes to

$$T(f) = \frac{a_{spine}(f)}{a_{seat}(f)}$$

where $a_{seat}(f)$ is the acceleration at the seat and $a_{spine}(f)$ is the acceleration at the spine at frequency f. So, a transmissibility of 2 would mean that there was twice as much vibration at the spine than at the driving point.

For measures of transmissibility, the investigator is usually interested in determining which frequencies resonate (i.e., show a peak in the transmissibility) and how damped is each resonance (the amplitude of the peak is inversely proportional to the damping). The investigator might also be interested in determining at which frequencies there is vibration isolation (i.e., there is less vibration at the remote point than at the driving point). If there is a peak in the transmissibility, then vibration at that frequency will be amplified by a buildup of stored energy in the repeated stretching and compression of tissue. At frequencies where isolation occurs, there is no buildup of stored energy.

Phase is an important concept when considering transmissibility of the body. If there was a transmissibility of 1 to a location on the body, and if the phase was 0° (i.e., in phase) then there would be no relative movement between the measuring points. For all other values of phase there is relative movement, even if the trans-

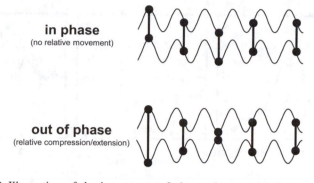

in phase
(no relative movement)

out of phase
(relative compression/extension)

FIGURE 2.19 Illustration of the importance of phase when considering two simultaneous vibration measurements. Although the signals have identical frequency and magnitude (and therefore a transmissibility of 1), the relative movement of material between the measurement locations depends on the relative phase of the signals.

missibility is 1. If, at some arbitrary frequency, there was a transmissibility of 1 to a location on the body but the phase was 180° (i.e., perfectly out of phase), then the relative movement between the measuring points would be maximized, indicating a cycling compression and extension (Figure 2.19).

Whole-body vibration transmissibility is usually measured to the spine or to the head. For measures to the lumbar spine, there are peaks at about 4 Hz and about 8 to 10 Hz that slightly reduce in frequency with increased vibration magnitude (Kitazaki, 1994; Mansfield and Griffin, 2000; Matsumoto and Griffin, 1998). This reduction in frequency is symptomatic of a nonlinear softening system, whereby the body loses stiffness as acceleration magnitude increases. This is likely to be due to a buckling of the spine under high acceleration, but this hypothesis is difficult to confirm using noninvasive methods.

Transmission of vibration to the head has been comprehensively investigated by Paddan and Griffin who measured translational (x-axis, y-axis, and z-axis) and rotational (roll, pitch, and yaw) responses of the head to a variety of stimuli (1988a, 1988b, 1993, 1994, and 1996). The importance of the seating condition was shown in these studies as the response of the head differed depending on the existence or otherwise of a backrest (Figure 2.20). For vertical vibration, there is a resonance in the transmission of seat vibration to the head (vertical) at about 4 to 5 Hz with no backrest contact. With a backrest, the resonance frequency is increased to 6 to 7 Hz. Vibration at 6 to 7 Hz also causes a pitch (nodding) movement of the head in combination with a fore–aft response, irrespective of backrest condition. This means that a single axis input (vertical) causes a complex response at the head.

Knowing the transmissibility allows for predictions of movement from measurements of vibration at the seat. For example, the visual performance of a tank driver might be compromised due to vibration at the head, which would be difficult or unsafe to measure in rough terrain. Using laboratory-based measurements of transmissibility combined with field measurements of vibration at the tank driver's seat, the extent of the degradation of performance could be estimated.

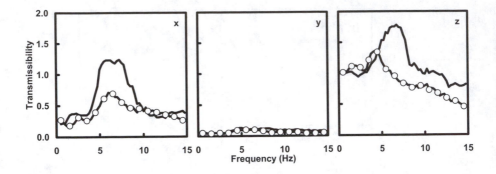

FIGURE 2.20 Median transmissibilities between vertical (z-axis) seat acceleration and head fore–aft (x-axis), lateral (y-axis), and vertical (z-axis) acceleration for 12 subjects seated with a backrest (——) and without a backrest (–○–). Data from Paddan and Griffin, 1993.

2.6.2 MECHANICAL IMPEDANCE OF THE BODY

The driving point mechanical impedance of the body is defined as the ratio of the force to the velocity at any frequency:

$$z(f) = \frac{F(f)}{v(f)}$$

where $z(f)$ is the driving point mechanical impedance and $F(f)$ and $v(f)$ are the force and the velocity at the driving point at frequency f.

An alternative method of expressing mechanical impedance is to use the apparent mass. Newton's second law states that:

$$F = m \times a$$

where F, m, and a are the force, mass, and acceleration, respectively. Adding a frequency term and rearranging, this equation becomes:

$$M(f) = \frac{F(f)}{a(f)}$$

where $M(f)$ is the apparent mass at frequency f. For rigid systems, the apparent mass is the same as the mass of the system. However, for compliant systems (such as the human body) under vibration, this equation is a function of frequency such that the apparent mass is greater at some frequencies than others. At very low frequencies, the apparent mass always tends to the total mass of the system. Therefore, if the apparent mass of a 50-kg person is measured and compared to the apparent mass of a 100-kg person, it can be observed that the apparent masses at 0 Hz will be 50 and 100 kg, respectively (Figure 2.21). To assist in comparing data from subjects of different weights, the apparent mass is usually normalized by the static weight,

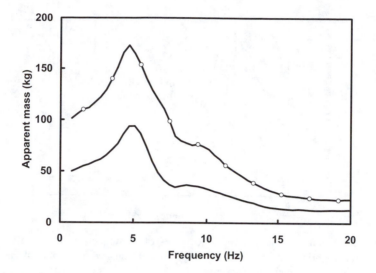

FIGURE 2.21 Apparent masses of a 100-kg subject (–o–) and a 50-kg (———) subject exposed to vertical vibration (non-normalized).

sometimes defined as the apparent mass at a fixed low frequency (e.g., 0.5 Hz). If the whole mass of the subject is supported on the force-measuring transducer, then the subject weight can be used for normalization, but often measures are taken of the force at the seat while the feet are resting on a moving, or fixed, platform. In these cases, the "sitting weight" should be used for normalization (i.e., the weight of the body supported on the seat).

Normalized apparent masses of seated subjects in the vertical direction show a clear peak at about 5 Hz and often a second peak at about 10 Hz (Figure 2.22; Fairley and Griffin, 1989; Mansfield and Griffin, 2000). These peaks reduce in frequency with increases in vibration magnitude, symptomatic of the body being a nonlinear softening system, as discussed in Subsection 2.3.1. The magnitude of the 5-Hz peak in the normalized apparent mass is usually about 1.5 to 1.7 but can reach 2 for some individuals. Therefore, at some frequencies, the load on a seat might be double of that which would be expected if the body were rigid. This is the reason why it is not appropriate to use rigid masses to represent the body when testing the dynamic performance of seats but, instead, human subjects or seat testing dummies should be utilized.

2.6.3 EFFECT OF SUBJECT VARIABLES

There is little evidence to support changes in the biomechanical response of the body to vibration with subject variables. The transmissibility to any location on the body shows some variability from one subject to another, but to date, there has been no convincing method of predicting these differences from anthropometric measurements. The only measure that changes with subject mass is the magnitude of the apparent mass, but this difference is eliminated once the measure is normalized. Indeed, Fairley and Griffin (1989) show that the mean normalized apparent mass of

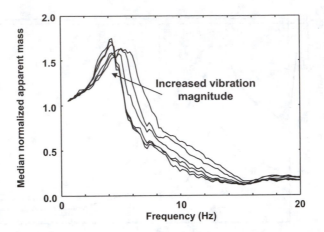

FIGURE 2.22 Median normalized apparent masses of 12 subjects measured at six magnitudes of vertical vibration from 0.25 to 2.5 m/s² r.m.s. Resonance frequencies decrease with increases in vibration magnitude (data from Mansfield and Griffin, 2000).

24 men was almost identical to the mean normalized apparent mass of 24 women and of 12 children.

Small changes of posture have also been shown to have only marginal effects on the apparent mass and rotational motion of the pelvis (Mansfield and Griffin, 2002). However, care should be taken in using data obtained using upright postures in applications where semisupine postures with steering wheel contact are used (e.g., driving a car) as there is some evidence to suggest that the resonance frequency increases and the magnitude of the peak decreases in such postures (Rakheja et al., 2002).

2.6.4 Effect of Stimulus Variables

The nonlinear softening characteristics of the apparent mass and transmissibility of the sitting person to vertical vibration have been described previously (Subsection 2.6.2). This means that measurements made at one magnitude of vibration will not necessarily be applicable at another magnitude.

Most of the experimental studies considered thus far in this section have used random vertical vibration as the stimulus. This type of stimulus has the advantage that all frequencies can be tested simultaneously, allowing for rapid generation of results. However, it is also possible to measure impedance and transmissibility at discrete frequencies using sinusoidal stimuli. If tests are made at a series of discrete frequencies, then a graph can be constructed of the results. Such data give results of a similar form to those measured with random signals (e.g., Holmlund and Lundström, 1998; Matsumoto and Griffin, 2002). Although sinusoidal signals are helpful in developing an understanding of the biomechanics of the seated body in laboratory conditions, they do not occur in real work environments.

Exposure to shocks is thought to be the most important part of a vibration signal for both comfort and health (e.g., Sandover, 1998). Using subjects with accelerom-

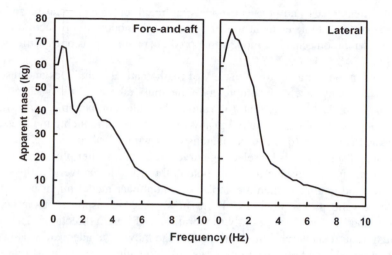

FIGURE 2.23 Apparent masses of seated subjects exposed to horizontal vibration measured by Fairley and Griffin (1989).

eters attached directly to their L4 spinous process and exposed to shocks generated by a pendulum impacting with a seat, Magnusson et al. (1993) show similar results for the transmissibility to the spine to those of Kitazaki (1994) who used random vibration. The apparent masses of subjects exposed to shocks and random vibration have been directly compared by Mansfield et al. (2001). Although slight differences were observed, the main features of the apparent mass were similar between conditions. This study also showed that the predictability of the shocks does not affect the apparent mass.

Exposure to whole-body vibration is rarely restricted to just vertical motion. Usually, there is some element of horizontal movement; often the horizontal axes can dominate. For example, Lundström and Lindberg (1983) reported multiaxis vibration magnitudes for 56 construction vehicles. For 13 of the vehicles there was more weighted vibration in one horizontal direction than in the vertical direction; for all others, the vibration magnitude in one horizontal direction was at least 90% of that reported for vertical motion. Therefore, it is important to consider the response of the body to nonvertical vibration. The apparent mass of the seated person exposed to fore-and-aft vibration shows peaks at about 0.7 and 2.5 Hz; in the lateral direction, peaks occur at about 0.7 and 2 Hz (Figure 2.23; Fairley and Griffin, 1990; Holmlund and Lundström, 1998). As for vertical vibration, the peaks in apparent mass in horizontal axes are nonlinear; i.e., the resonance frequencies decrease with increases in vibration magnitude (Mansfield and Lundström, 1999b).

2.6.5 BIOMECHANICAL MODELS FOR HUMAN RESPONSE TO WHOLE-BODY VIBRATION

Biomechanical models for application to whole-body vibration can take a variety of forms (Griffin, 2001). For whole-body vibration, the most common types of models are lumped parameter and physical models, although other types can be found in the literature (e.g., finite element models, continuous models). This section is only concerned with biomechanical models that predict apparent mass or transmissibility of the body.

Models in the literature use a variety of mathematical techniques and result in models with many levels of complexity. One must take care when selecting or developing a model to ensure that it is valid and usable. For example, it is possible to fit a complex mathematical model to any simple system producing unnecessarily complicated equations (this can be demonstrated with a few mouse operations in most standard spreadsheets by selecting curve-fitting in a scatter plot). In this case, the model might be unusable for some potential users. Alternatively, if the system is complex and nonlinear, then a complex and nonlinear model might be required for it to be valid. It has previously been shown that the response of the body is nonlinear and it responds differently to vibration in different directions. Therefore, one might require a complex model to predict the movement adequately. However, it is often possible to break a complex system into simpler components that serve the purpose just as well. For example, if researchers are interested in the response of a generalized (or even standardized) person on a seat exposed to vertical vibration, then they might be justified in using the average response of a number of individuals and neglecting the nonlinearity. This model will not be representative of any single individual, but will represent their average response. An alternative approach is to design for populations by considering each of the individuals (or groups of individuals) in that population. For example, separate models could be developed for 5th, 50th, and 95th percentile male and female children, adults, and elderly people. For the purposes of human vibration applications, it has been shown that the only major differences between subjects are their body weights; therefore, similar models could be used for these different groups, but the only need is to scale the parameters for the various weights. One challenge for the development of all models is to ensure that the data on which it is based is of high quality. If there are doubts over the validity of the source data, then there will be doubts over the model.

2.6.5.1 Lumped-Parameter Models

For whole-body vibration, the most common form of mathematical model in the literature is the lumped-parameter model. This type of model is built from masses, springs, and dampers. The values of the mass, stiffness, and damping are selected such that the response of the lumped-parameter model represents the response of the person. It is important to note that it is not possible to map values of stiffness or damping provided in lumped-parameter models onto any specific biological tissue in the human. The simplest form of lumped-parameter model is a single degree-of-freedom system that is made from a single moving mass supported on a single spring and damper (Figure 2.24). A single degree-of-freedom system has a single peak in the response. If there is no damping, then the peak occurs at:

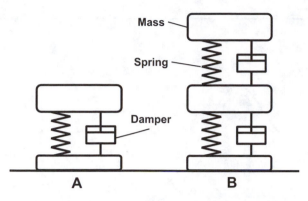

FIGURE 2.24 Simple lumped-parameter models consisting of masses, springs, and dampers. A: single degree-of-freedom system; B: two degree-of-freedom system.

$$f_n = \frac{1}{2\pi} \times \sqrt{\frac{k}{m}}$$

where f_n is the natural frequency, k is the spring stiffness, and m is the moving mass of the system. If there is damping in the system, then the resonance frequency occurs at a slightly lower frequency related to the viscous damping ratio, ζ:

$$\zeta = \frac{c}{2\sqrt{km}}$$

where c is the damping constant of the damper. The damped natural frequency, f_d, is:

$$f_d = f_n \sqrt{\left(1 - \zeta^2\right)}$$

More complex models can be built using the same components in series or parallel. Increasing the number of moving masses in a system will increase the number of peaks present in the response (for a more comprehensive discussion of these topics, see Lalor, 1998).

For vertical apparent-mass models, single degree-of-freedom models are usually sufficient (Figure 2.25; Fairley and Griffin, 1989). The second peak observed for individual subjects can be reproduced if the model is extended to two degrees-of-freedom (Wei and Griffin, 1998). Three mass models are required to represent the apparent mass of seated subjects exposed to horizontal vibration (Mansfield and Lundström, 1999a). Although lumped-parameter models are "constructed" from masses, springs, and dampers, they only exist as mathematical simulations. However, they can still be used with measures of the dynamic properties of seats to predict their dynamic performance.

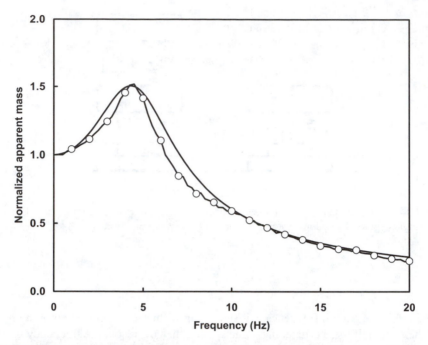

FIGURE 2.25 Mean normalized apparent mass of 60 people (–○–) compared with the response of a single degree-of-freedom lumped-parameter model (————). [Adapted from Fairley, T.E. and Griffin, M.J. (1989). The apparent mass of the seated human body: vertical vibration. *Journal of Biomechanics,* 22(2), 81–94.]

2.6.5.2 Physical Models

A second type of the biomechanical response model of the seated person is a physical model, sometimes known as a seat-test dummy or anthropodynamic dummy (see also Subsection 2.5.4). Such devices can be physical representations of lumped-parameter models already discussed in this section and are not intended to look like people. The main challenges with anthropodynamic dummies are in the engineering and practicalities of the design. Lumped-parameter modeling assumes linear and frictionless components in the system. In practice, dampers are usually nonlinear, have some friction, and change their properties with temperature. Lumped-parameter models also assume uniaxial vibration and only represent the response of the body in that direction. Even if a seat is tested in the laboratory with only vertical vibration on a single-axis shaker, the dummy will move in more than one direction (albeit still dominated by its vertical response). In a moving vehicle there is vibration in all axes simultaneously, and a single-axis dummy will not move like a human occupant. A final challenge with anthropodynamic dummies is that of the health and safety of the test team. For it to be representative, the dummy must have a mass similar to that of a person. Therefore, it poses a manual handling problem, especially when installing or removing the device from a space-limited cabin such as that of a car. Also, the dummy must be safe in all driving situations, including emergency procedures. If the vehicle must suddenly stop to avoid a collision, the dummy must remain securely tethered to the seat.

Notwithstanding the problems associated with anthropodynamic dummies, a range of devices have been developed (e.g., dummies developed at the University of Southampton, U.K.; Darmstadt University of Technology, Germany; University of Vermont, U.S.; Figure 2.26). Most are constructed from masses, springs, and dampers, although some have been developed that use active control algorithms to represent subjects of different physical characteristics or even the nonlinearity in apparent mass with vibration magnitude (Lewis and Griffin, 2002). Seat-test dummies have been proven to give representative measures of seat transmissibility and therefore may become the standard devices used by industry to test conventional and suspension seats. It would be of mutual benefit to standardize dummy perfor-

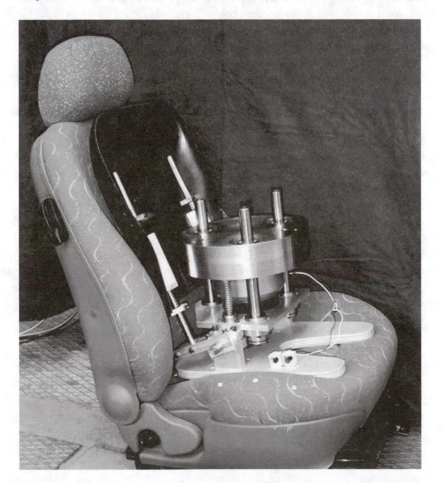

FIGURE 2.26 Prototype "Memosik" anthropodynamic dummy mounted on a suspension seat (Cullmann and Wölfel, 2001). (Image © Darmstadt University of Technology, used with permission.)

mance such that all meet a minimum specification. Work is underway through the International Organization for Standardization to specify dummy performance for application to mobile machines. It is possible that dummies that meet the specification might also be appropriate for use in other forms of transport, but one should remain cautious in using a device in an environment for which it is not intended.

2.7 CHAPTER SUMMARY

Usually, whole-body vibration is associated with traveling in vehicles and it can affect comfort, performance, and health. The body is most responsive to vertical whole-body vibration at about 5 Hz, which corresponds to peaks in the vertical frequency weightings (W_b and W_k) and in the apparent mass. In horizontal axes, the body is most responsive at frequencies below 2 Hz, which also corresponds to peaks in the horizontal frequency weighting (W_d) and in the apparent mass. At low magnitudes of vibration, the primary concern is with comfort, with activity interference, and with vehicle refinement. At higher magnitudes, vibration constitutes a health risk, and occupations with high vibration exposures have been linked with high incidences of back pain. In many industries, occupational vibration exposures should be reduced to minimize the risk of injury. One method of reducing exposure is to design seats such that they isolate the occupant from the vibration. Conventional foam seats, such as car seats, have a resonance at 4 to 5 Hz, which coincides with the frequency where the body is most responsive to vibration. Suspension seats have a resonance frequency at about 2 Hz and are appropriate in many vehicles where vibration constitutes a health risk. Care must be taken in the specification and adjustment of suspension seats to ensure that occupants are not exposed to end-stop impacts that might introduce an even more serious risk. Biomechanical investigations of the response of the body to whole-body vibration have included measures of transmissibility and apparent mass. These data have provided tools for prediction of the vibration at the head and spine from measurements at the seat and have also enabled the development of mathematical and physical models of the response of the whole body to vibration.

3 Motion Sickness

3.1 INTRODUCTION

Seasickness, coach sickness, car sickness, airsickness, or even space sickness are phenomena that have been experienced by most travelers. These can be broadly categorized as different forms of motion sickness. One does not even need to move to experience motion sickness; one common side effect of virtual reality systems or playing computer games with large monitors or headsets is nausea. All environments where such sickness occurs have the common element of real or apparent motion (Table 3.1).

The term *sickness* implies a problem of a medical nature, but motion sickness is a normal and natural response to some forms of movement. Its ultimate effects might lead to a requirement for treatment (e.g., for fluid loss following vomiting), but motion sickness in itself is not directly fatal. Sanders and McCormick (1993) state: "Although motion sickness may never actually kill us, there are times when we wish it would!"

This statement highlights one particular problem that is frequently associated with motion sickness: a state of apathy, impaired performance, or both which could result in an increased probability of misjudgment.

The common root of the words "nautical" and "nausea" indicates the long association of motion sickness with travel in general, and sea travel in particular. (Both words are derived from the Greek word "naus" meaning "ship.") For centuries the experience of motion sickness was restricted to sailors, riders of camels and elephants, and to those traveling by horse-drawn coaches. As the available modes of travel have increased since the 19th century, the number of situations where sickness might occur has also increased. Almost all modes of transport have some association with motion sickness, and as each of these has become more affordable and available to the general public, the incidence of travel sickness has increased.

Modern scientific observation of motion sickness symptomatology can be dated back to at least the 18th century with Erasmus Darwin's *Zoonomia* (1796).* In addition to observing that individuals can become "accustomed" to riding on elephants, he reported from his own observations:

> In an open boat passing from Leith to Kinghorn in Scotland a sudden change in the wind shook the undistended sail and stopped our boat; from this unusual movement the passengers all vomited except myself. I observed that the undulation of the ship

---

* Erasmus Darwin was the grandfather of Charles Darwin, although Erasmus had died prior to Charles' birth.

TABLE 3.1
Some Examples of Environments Where
Motion Sickness Might Occur

Sickness Associated with Mode of Transport
Cars
Large sea vessels (e.g., ferries, navy ships)
Small sea vessels (e.g., yachts, small boats)
Coaches (buses)
Coaches (horse-drawn)
Aircraft
Space travel
Trains
Camel or Elephant rides (but not horses)

Sickness Associated with a Moving Visual Scene
Virtual reality
Film/video editing
Microfiche readers
Microfilm readers
Simulators (fixed base)
Computer games
Cinema

Other Situations Where Motion Sickness Might Occur
Coriolis stimulation (e.g., merry-go-rounds)
Barbecue-spit rotation
Sea swimming
Wearing distorting spectacles
Fairground rides
Simulators (moving base)

---

and the instability of all visible objects inclined me strongly to be sick; and this continued or increased when I closed my eyes, but as often as I bent my attention with energy on the management and mechanism of the ropes and sails, the sickness ceased; and recurred again as often as I relaxed this attention; and I am assured by a gentleman of observation and veracity that he has more than once observed when the vessel has been in immediate danger that the seasickness of the passengers has instantaneously ceased and recurred again when the danger was over.

This observation highlights the two elements of motion sickness: the physiological response to the motion (i.e., the response to the undulation and visual instability) and the psychological response (i.e., the ability to reduce the symptoms through distraction).

An individual exposed to a nauseogenic stimulus (i.e., a stimulus that induces motion sickness) might experience a variety of symptoms (Section 3.2) caused by a combination of physiological and psychological responses (Section 3.3). The symptoms could occur in a real (Section 3.4) or virtual environment (Section 3.5). Although almost everybody can be made to feel sick if a stimulus is generated at the most provocative frequency (Section 3.6), individuals might be able to habituate themselves (Section 3.7) or take other preventative measures (Section 3.8) to minimize their adverse responses.

3.2 SIGNS AND SYMPTOMS OF MOTION SICKNESS

The most tangible result of motion sickness is vomiting, but this is not the only symptom. Other symptoms include dizziness, bodily warmth, sweating, drowsiness, yawning, loss of appetite, increased or decreased salivation, headache, lethargy, "stomach awareness," burping, nausea, and pallor. These symptoms need not be severe enough to cause discomfort and distress to the traveler. Using quantitative methods, Holmes et al. (2002) show that an individual's skin color changes while experiencing motion sickness (possibly indicating that describing an affected individual as "turning green" might be accurate).

Usually the signs and symptoms of sickness progress through a sequence, although the precise sequence content, order, and the speed at which symptoms develop depends on the individual. Generally, yawning, bodily warmth, and stomach awareness develop first. Following this, there may be a change in mouth dryness (either more salivation or a drying of the mouth), as well as initiation and development of an intense feeling of nausea combined with a feeling of apathy. Finally, vomiting might occur. Unfortunately, the act of vomiting does not guarantee an end of the suffering, as although vomiting may provide temporary relief, the symptoms and signs can reoccur, resulting in repeated vomiting until the motion stimulus ceases or the individual is removed from the environment.

The first symptoms may take some hours to develop but can arise within minutes. When the symptoms become more severe, an "avalanche phenomenon" or "cascade effect" usually occurs whereby the later symptoms develop sequentially and rapidly with an inevitable end point (vomiting) if the stimulus is maintained (Reason and Brandt, 1975; Brandt, 1999).

One of the most dangerous situations in which motion sickness can occur is for those isolated at sea, as in a life raft. Although the occupants might have been accustomed to the movement of the larger vessel that was abandoned, the different nature of the motion of the smaller boat can initiate symptoms of sickness. For those suffering with seasickness in a life raft, repeated vomiting can cause serious dehydration, especially when fresh water supplies are limited. If the dehydration is coupled with the sense of hopelessness which usually accompanies the situation, the chances of survival can be significantly reduced.

3.3 THE CAUSE OF MOTION SICKNESS

Motion sickness provides no benefit to the sufferer and has not been shown to provide some evolutionary advantage. Indeed, in some situations, the sickness might actually jeopardize chances of survival. A question therefore arises: why do some forms of motion make us sick?

Motion sickness had been attributed to a variety of theories before the theory of sensory conflict was accepted. These were broadly categorized as psychological and physiological. The psychological theories were based on ideas of fear of travel or fear of the prevailing conditions. The physiological theories were based on ideas of overstimulation of the stomach (or other organs) due to the individual being exposed to too much movement. Considering the relatively limited knowledge of medical science at the time, these theories were logical but failed to apply to all types of sickness.

The currently accepted theory combines physiological and psychological components and is known as "sensory conflict" or "sensory rearrangement" theory.

3.3.1 SENSORY CONFLICT THEORY

Motion sickness, by definition, occurs when an individual is exposed to real or apparent motion. The types of motion that are nauseogenic are low-frequency, high-displacement stimuli. Such stimuli are sensed by the somatic, visual, and vestibular systems (also see Subsection 2.2.1 for a description of the mechanisms of motion perception). In many situations the stimulus is controlled by the individual (e.g., driving a car), and so an extra pseudosensory input of "control" can be added to the list of "senses."

Usually, the brain simultaneously receives a battery of motion-sensing signals from these four systems (somatic, visual, vestibular, and control). For example, while turning a corner in a car, the visual system will perceive objects moving to the left or right, the semicircular canals of the vestibular system will sense the rotational movement, the otoliths in the saccule and utricle will respond to the centrifugal forces, the somatic system will feel the changes in pressure across the body, and the driver will know that the process of cornering is happening because the steering wheel had been previously turned (Figure 3.1a). This provides a consistent cognitive model of the motion environment whereby all systems are providing coherent information to the brain.

In some situations, the motion-sensing signals being integrated by the brain are inconsistent with each other. For example, a passenger in a car might be reading a newspaper while the driver steers the vehicle around a corner. In this case because the passenger is not controlling the steering wheel or concentrating on the route, the turn probably will not be anticipated, and the visual system will not see the moving visual scene but only the words on the page, which move with the car and therefore appear to be stationary. However, the semicircular canals will sense the rotational movement, the otoliths will respond to the centrifugal forces, and the somatic system will feel the changes in pressure across the body (Figure 3.1b). This provides a cognitive model of the motion environment that is internally inconsistent. The visual system is in conflict with the other senses, and the lack of control input fails to reinforce the correct cognitive model. It is interesting to note that in contrast to the response of drivers who almost never get sick, it is common for individuals to report that reading in cars or buses is nauseogenic (Probst et al., 1982).

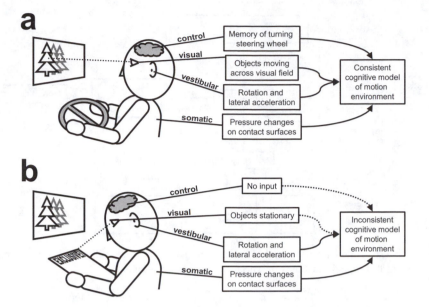

FIGURE 3.1 Illustrative model of the multiple pathways through which a motion environment is perceived: (a) for a driver of a car and (b) for a passenger in the same car reading a newspaper.

Although this simple model of the sensory response to motion is helpful in understanding the underlying mechanisms of motion sickness, it does not explain how individuals are able to habituate to motion that initially caused sickness. Furthermore, there are phenomena where, after becoming habituated to a nauseogenic stimulus, the individual will experience a return of sickness when the stimulus is removed (this is termed *mal de debarquement* by sailors).

The basis for a theory on the cause of motion sickness is that if there is a conflict between the expected and the actually experienced sensory signals, a sense of imbalance can result in sickness. This is known as the *sensory conflict* or *sensory rearrangement theory* (Figure 3.2). If the experienced combination of signals matches the expected combination there is no change in physiological state and homeostasis is maintained. If the perceived combination of signals does not match the expected combination, then sensory rearrangement is detected and the symptoms of motion sickness might appear. If a novel combination of signals is experienced (i.e., a combination which is not found in the memory of expected combinations of sensory inputs) then the memory is updated. Hence, with repeated exposures to a conflicting combination of sensory signals, the likelihood of matching with a combination in memory is increased. Factors that might increase or decrease the severity of sickness symptoms include the sex of the individual exposed, whether drugs have been consumed (whether the drugs are related to motion sickness or not), other aspects of the physical environment (such as air temperature and air quality), and the psychological state of the exposed person.

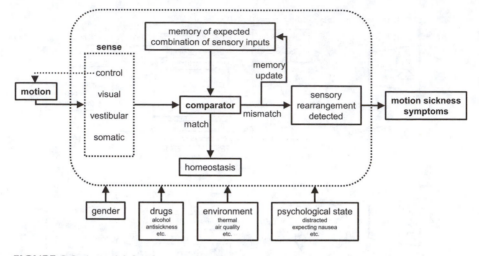

FIGURE 3.2 A model for the genesis of motion sickness that complies with sensory rearrangement theory and illustrates the mechanism of habituation.

In all types of motion sickness a sensory conflict can be identified although sensory conflict need not necessarily result in sickness. In most types of land-travel sickness, the visual system provides incorrect information to the brain due to its being focused on the interior of the vehicle or on an item that moves with the interior of the vehicle, whereas the vestibular and somatic systems respond to the movement of the vehicle itself (Figure 3.3a). For seasickness the visual system might be focused inside a cabin, on the sails, on a chart, or on displays from navigational aids. In all these cases the view moves with the boat, providing a visual stimulus which is stationary. Meanwhile, the movement of the boat is sensed through other balance systems. For most forms of simulator sickness the situation is reversed. The visual system (and control system) provides signals indicating that motion is occurring, whereas the vestibular and somatic systems provide the correct signals that motion is not occurring. In simulators where the movement is in response to the operator's actions, the movement is usually an imperfect recreation of reality, and so a conflict still occurs (Figure 3.3b).

Although it is possible to identify the mechanisms of sensory conflict in all provocative situations, it is less apparent as to why a conflict in sensory signals leads to sickness. To develop an appreciation of why motion might make us sick, it is necessary to understand the principles of some of the mechanisms that are triggered by the ingestion of certain toxins.

Many toxins affect the nervous system directly or indirectly after being consumed. One type of toxin which might have been experienced by some readers is alcohol and so this will be used as an example; however, many other toxins affect the nervous system in a similar manner (e.g., opiates such as morphine). If alcohol is consumed in sufficient quantities, then permanent damage or even death can result, therefore, it can be seen how functional it is for the body to naturally react with emesis (i.e., vomiting) as a form of self-protection. Absorption of the alcohol

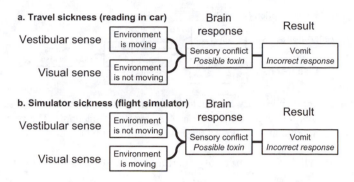

FIGURE 3.3 Examples of sensory conflict between the visual and vestibular senses, the brain responses, and ultimate results for (a) most forms of travel sickness (e.g., reading in a car), and (b) most forms of simulator sickness (e.g., fixed-base flight simulator) for nonhabituated individuals. For both types of sickness the brain response and result are incorrect.

by the blood causes a slight reduction in blood density. As the cupula has its own blood supply, its density is also reduced (due to the change in blood density), and this causes it to become buoyant in the endolymph (see Subsection 2.2.1). Therefore, the cupula loses stability, and this results in an upward pressure on the embedded hair cells which connect it to the semicircular canal wall. Such forces are identical to those that would result if the environment was actually rotating. The hair cells at the bases of the cupulae are distorted and send nervous signals to the brain. Therefore, a feeling of vertigo will be experienced by the person drinking alcohol, who commonly describes the effect as "the room was spinning." A sense of dizziness is experienced after ingesting many naturally occurring toxins. Therefore, it is appropriate for the brain to induce emesis if it perceives that a possible toxin has been consumed.

   Considering an evolutionary timescale, it was once far more likely for an individual to consume food containing a toxin than to be exposed to passive low-frequency stimuli. Therefore, if the brain perceived a low-frequency oscillation or continuous rotation without an expected "control" or visual input (i.e., a sensory conflict occurred), then it would be potentially lifesaving to vomit (Triesman, 1977; Money and Cheung, 1983; Money, 1990; Figure 3.4). In modern times, the likelihood of exposure to low-frequency motion in transportation has increased, while food hygiene standards have also improved, thereby decreasing the likelihood of ingesting a toxin. Therefore, the probability of sensory conflict being potentially fatal has substantially declined in recent centuries (at least for individuals who do not abuse alcohol!). However, the legacy of our evolutionary safety valve remains.

3.3.2 Types of Sensory Conflict

Most sensory signal conflicts can be categorized into two groups (Benson, 1984): visual–vestibular mismatches and intravestibular mismatches. Within the two categories of conflict, either both sensory systems give inconsistent signals or one system provides a signal and the other does not. These are known, respectively, as Type I

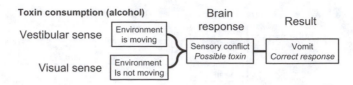

FIGURE 3.4 Example of sensory conflict between the visual and vestibular sense, the brain response, and results following consumption of a toxin (e.g., alcohol). The brain response and result are an appropriate reaction to the ingestion of the toxin.

and Type II mismatches. Examples of situations where such conflicts occur are listed in Table 3.2.

Visual–vestibular mismatches are the most common forms of sensory conflicts. Fortunately, these are also the easiest to design out of a system or for the affected person to control. For example, Type I visual–vestibular mismatches can be alleviated by ensuring that the visual system provides signals that are coherent with the vestibular system. For Type II mismatches, either the stimulus can be terminated or a coherent combination of signals can be sought (e.g., by obtaining an external view in a vehicle).

An intravestibular mismatch occurs when the semicircular canals and the otoliths provide contradictory or unrelated signals. Although these are relatively rare, they tend to be more difficult to eliminate and the sickness cannot be alleviated by behaviorable changes. For example, in space sickness, fluids in the semicircular canals retain their inertia, generating rotation signals. However, due to the absence of gravity, the expected corresponding otolith signals are not generated when the head pitches or rolls.

Individuals whose vestibular systems do not function do not develop the symptoms of motion sickness, even for those environments where there is no vestibular input (i.e., visually induced sickness). This is because the combination of signals received by the brain from the senses is coherent with the neural store; visual and somatic responses are never experienced in combination with a vestibular response.

3.4   TRAVEL SICKNESS

Traveling in a vehicle is the most widespread situation where motion sickness occurs. The majority of people have experienced travel sickness in at least one mode of transport.

3.4.1   SEASICKNESS

Seasickness can occur on all types of vessels, from life rafts to yachts, from luxury cruise ships to hovercrafts, from lifeboats to fishing boats. The visual–vestibular conflict that occurs is either Type I, if the traveler is on deck or watching a moving scene (e.g., a sail), or Type IIb, if the traveler is below the deck or viewing a fixed visual scene (e.g., a book).

Lawther and Griffin (1988a, 1988b) conducted a questionnaire survey of motion sickness symptoms among passengers of 114 ferry voyages. Of the 20,029 passen-

TABLE 3.2
Examples of Categories and Types of Sensory Conflict

	Category of Motion Cue Mismatch	
	Visual–vestibular (Visual, A; Vestibular, B)	Intravestibular (Cupula, A; Otolith, B)
TYPE I **A and B simultaneously give contradictory or uncorrelated information**	Watching waves from side of ship Watching sail on a boat Watching passing cars on a coach Use of binoculars in a moving vehicle Moving-base flight simulator	Making head movements while rotating (coriolis or cross-coupled stimulation) Making head movements in an abnormal force environment which may be constant (e.g., hyper- or hypogravity) or fluctuating (e.g., linear oscillation) Space sickness Vestibular disorders (e.g., Ménière's disease)
TYPE IIa **A signals in absence of expected B signals**	Computer game sickness Cinerama sickness Fixed-base simulator sickness Circular vection	Cornering in a tilting train Positional alcohol nystagmus Caloric stimulation of semicircular canals Vestibular disorders (e.g., pressure vertigo, cupulolithiasis)
TYPE IIb **B signals in absence of expected A signals**	Looking inside a moving vehicle without external visual reference (e.g., below deck in boat) Reading in a moving vehicle	Low-frequency (< 0.5 Hz) linear oscillation Rotating linear acceleration vector (e.g., barbecue-spit rotation, rotation about an off-vertical axis)

Source: Adapted from Benson, A.J. (1984). Motion sickness. In *Vertigo*, Dix, M.R. and Hood, J.D., Eds. New York: John Wiley & Sons.

gers who completed the questionnaire, 1,404 vomited (7%) and 5,939 (30%) felt "slightly unwell" or worse. For passengers traveling by hovercraft, lower incidences of sickness occurred, even when the effects of shorter crossing durations were taken into account. The ship's movement was dominated by a complex motion at about 0.2 Hz, irrespective of the type of ship or crossing (although the amplitude of the motion was dependent on the sea state). The hovercraft's motion did not have such a clear peak in energy within any single frequency band but showed more motion at higher frequencies. These findings are in agreement with laboratory studies that have shown a peak in the nauseogenicity of motion at about 0.2 Hz (e.g., O'Hanlon and McCauley, 1974; see Section 3.6) which unfortunately coincides with the peak in motion experienced on passenger ferries.

Seasickness is common among sailors of yachts. In a survey of the 182 partic-ipants in the British Steel Challenge round-the-world yacht race for inexperienced

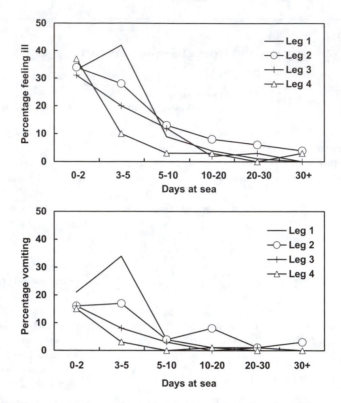

FIGURE 3.5 Illness occurrence and vomiting occurrence as a function of the number of days at sea and the race leg for sailors competing in a round-the-world yacht race (data from Turner and Griffin, 1995).

crews, Turner and Griffin (1995) report that almost half of the sailors vomited at some time during the race. For each of the four legs of the race, motion sickness symptoms reduced substantially after 5 days at sea (Figure 3.5). The extended period of sickness in Leg 2 was due to the heavy seas experienced by the crews in the Southern Ocean and around Cape Horn. The data show that although the participants' tolerance to the provocative stimulus increased within each leg, the increased tolerance wore off by the start of the subsequent leg. This is consistent with anecdotal reports that it can take some days for yacht crew members to become reaccustomed to the movement of the boat when returning to the ocean following a period on land.

Perhaps the earliest measure of human vibration exposure is sea state, and this has been shown to correlate with sickness incidence (Turner and Griffin's subjects reported no illness below the category called Sea State 2). Other suggested methods of assessing the severity of the provocative stimulus for seasickness have occasionally included liters or meters (!), but the only validated measure is the motion sickness dose value (MSDV). The MSDV is defined in British Standard 6841 (British Standards Institution, 1987) as:

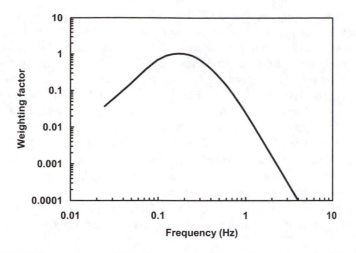

FIGURE 3.6 W_f frequency-weighting curve for prediction of seasickness incidence on passenger ferries as used by BS6841 (1987).

$$MSDV = \sqrt{a_w^2 t}$$

where a_w is the r.m.s. of the frequency-weighted acceleration and t is the duration of the motion exposure. The frequency weighting to use for motion sickness is W_f which has a peak of 0.2 Hz (Figure 3.6). One simple interpretation of the MSDV is the estimate of:

% vomiting = 1/3 MSDV

The MSDV is only recommended for use with vertical motion and has been designed for application on passenger ferries. However, it is possible that it could be used in other modes of transport to predict incidence of illness, if the motion is similar in nature to that of a ferry.

3.4.2 AIRSICKNESS

In all types of aircraft (e.g., cargo, passenger, helicopter, glider, fighter) with all types of occupants (e.g., military, civilian), airsickness can be a problem. Indeed, much of the effort that has gone into motion sickness research has been to benefit military aviators. The sensory conflict in aviation is usually either visual–vestibular Type IIb (for passengers inside a cabin) or intravestibular (canal–otolith) Type I (for high g-force operations).

Improvements in large, long-haul passenger aircraft have enabled routes to be planned which avoid turbulence, and so the problem for civilians has been reduced. However, with the increased demand in short-haul "city hopper" flights which fly at lower altitudes and have less scope for rerouting, it is possible that airsickness will again become more of an issue in the future. For short-haul flights, symptoms rarely progress to vomiting, but the feeling of nausea is experienced by about 10% of all passengers, and symptoms are correlated to the MSDV (Turner et al., 2000).

Airsickness among trainee military pilots and navigators is common, with reported incidences ranging from 11% to 83% (Bagshaw and Stott, 1985; Carr, 2001). Low-frequency, high-acceleration maneuvers are an inherent part of flying fast military jets, and so it is inevitable that a susceptible trainee will experience motion sickness. Most trainee pilots quickly habituate to the motion but a residual 15% find the sickness compromising their performance and require assistance, initially with the use of antisickness drugs. One of the problems with these drugs is their side effects that include drowsiness; therefore, they are not suitable for single-seater aircraft. An alternative strategy for combating airsickness is the implementation of a desensitization program (see Subsection 3.8.6). Considering the expense of training military aircrew, it is more cost effective to persevere by forcing habituation than to terminate training prematurely and then re-recruit.

3.4.3 SICKNESS ASSOCIATED WITH GROUND TRANSPORTATION

In contrast to air and sea travel where the motion includes vertical low-frequency components, ground transportation is dominated by lateral acceleration at low frequencies. Apart from occasional road features (e.g., humpback bridges), low-frequency vibration is caused by acceleration, deceleration, and cornering. The conflict is usually caused by passengers in the vehicle not maintaining an external view (visual–vestibular Type IIb) or by passengers watching passing vehicles (visual–vestibular Type I).

In a survey of over 3000 coach (i.e., long-distance bus) passengers, Turner and Griffin (1999a) reported that more than 25% felt slightly unwell or worse during their journeys. The feeling of sickness was more common for those who did not have a clear view of the road. Among those who had no external view, 3.4% reported vomiting compared to less than 1% for those who had a good external view. Sadly, it is impossible to provide a clear view of the road ahead for all, but passengers should be advised to occupy those seats with the best view possible and designers should avoid placing unnecessary obstacles in the field of view.

The style in which the vehicle is driven is a key component in how provocative road vehicle motion might be. As the primary low-frequency stimuli are directly controlled by the driver (acceleration, deceleration, and cornering speed), the driver can reduce (or increase) the nauseogenicity of the motion by driving with different levels of aggression. For example, the coach driver's driving style can influence the low-frequency acceleration in horizontal axes more than the differences in horizontal acceleration observed between different coach types (Turner and Griffin, 1999b). Therefore, road vehicle drivers are able to influence how sick their passengers will become. To minimize the nauseogenicity of the journey, drivers should try to reduce acceleration and deceleration by anticipating junctions and slowing gradually over a greater distance. When starting from stationary, drivers should increase speed gradually with smooth changes of gear. Drivers should also try to reduce speed before cornering. Any journey time lost by driving slowly on urban roads with many junctions and turns can be made up on straight roads (e.g., motorways) without compromising the well-being of the passengers. Alternatively, one could consider the elimination of the need to take stops for sick passengers as a time saver.

Although trains have historically been considered as being only a mildly provocative stimulus for motion sickness, the extending high-speed rail network running tilting trains has caused this view to be challenged (e.g., Förstberg and Kufver, 2001). With tilting trains, there is a Type IIb visual–vestibular conflict, as well as a Type I intravestibular conflict, as the inertial force acting on the otoliths is fully or partially compensated by the tilt of the train. Therefore, there is a more complex combination of potentially provocative motions than might otherwise be assumed.

3.5 SIMULATOR SICKNESS

Simulators can take a variety of forms. Although aircraft simulators are often the first type that comes to mind, they are also used for many other types of military applications including armored fighting vehicles, helicopters, and ships. Simulators are also used for civilian applications ranging from transportation (aircraft and ships) to medical applications, through to providing a virtual model of a new structure (e.g., an oil rig or a building).

There is a range of ways in which a simulator can be used. For training applications, they are used not only for giving novices their initial exposure to what usually is an expensive piece of hardware (e.g., military or civilian aircraft and spacecraft), but also to enable simulation of emergencies for experienced pilots or drivers that would be impractical to reproduce on a real platform (e.g., engine failure at critical times, such as during takeoff for a passenger jet). In these situations, the response of the simulator must mimic the response of the real platform as closely as possible. Another form of training for military personnel is the virtual battlefield where the strategic approach is the focus of the simulation and the fidelity of the response of each individual vehicle is less critical. In these simulations, large numbers of military personnel are able to fight against or alongside each other by occupying mock-ups of various types of combat vehicles connected through a central controlling computer. A final application for simulators occurs when the controls of a remote device are operated through a computer link. For example, this could be used in an unmanned combat air vehicle (UCAV) or in remote or robot-assisted surgery. For these applications, feedback to the operator is required so that natural interaction is possible with the remote device. For each of these applications, there is a variation in the requirement for realism of the simulation. However, in all cases, there exists a form of sensory conflict, as there is always a limit to the performance of the hardware.

Most simulators have some form of visual display, whereas some have auditory components and others have moving parts. The visual display could be a virtual-reality (VR) headset, a wrap-around projection system, simply a computer monitor, or an array of screens. Most have some element of computer control and use computer graphics software to create the visual scene, although some flight simulators use a miniature video camera which flies across a miniature model terrain. The most expensive simulators are mounted on moving bases in such a way that movements seen on the screen are also sensed through the vestibular and somatic systems. It is impossible to simulate all sensations with a moving-base simulator: for example vehicle acceleration is imitated by tilting the simulator such that the otoliths are displaced appropriately. However, the rotation also causes a response from the semicircular canals which does not occur in reality, and hence an intravestibular conflict occurs. The type of conflict experienced in simulators is either visual–vestibular Type IIa (for fixed-base simulators) or either visual–vestibular Type I or intravestibular Type I (for moving-base simulators) or both.

Simulator sickness has been reported in the literature of many applications including tanks (Lerman et al., 1993), aircraft (Kennedy et al., 1989), and helicopters (Crowley, 1987). Kennedy et al. (1989) report that, for the U.S. Navy, the most nauseogenic simulators are helicopter simulators with a moving base and those with a wide field of view.

Within VR research and development, the idea of "presence" is central. To have presence, the user has to have a feeling of being there; i.e., the user is fooled into believing that the virtual environment is indeed reality and interacts with it accordingly. Presence is enhanced by increasing the field of view or increasing the fidelity of the image. If there is no sense of presence then conflict is less likely to occur, as the brain does not have an expected combination of sensory inputs with which to compare. Paradoxically, as engineers improve VR systems to provide closer representations of reality, the likelihood of (and the intensity of) sensory conflicts increase and the users are more likely to feel sick.

A second paradox with simulator sickness is that novice operators of many vehicles (especially when simulators are used for training) experience symptoms of sickness in the real vehicle environment. Therefore, one could argue that a perfect simulation of the real environment should also produce similar degrees of sickness. Perhaps the ideal simulator will be one in which experienced operators do not feel sick but trainees experience (and are able to habituate to) symptoms identical to those in the real vehicle. Unfortunately, the converse is true: experienced pilots tend to show more symptoms than novices, due to there being a more established expected combination of sensory inputs that are in conflict with the experienced combination of sensory inputs (Kennedy et al., 1990).

3.6 RELATIVE SENSITIVITY TO SIMULATED MOTION OF DIFFERENT FREQUENCIES

Testing the relative nauseogenicity of low-frequency vibration at various frequencies is practically difficult to achieve in the laboratory. High-displacement motion simulators are required to reproduce the test stimuli, and there are very few such facilities available even globally. Low-frequency vertical motion is particularly difficult to simulate, as the shaker must be capable of accelerating the load against gravity, as well as overcoming its inertia. In the horizontal direction, there is no change in elevation, and therefore no change in potential energy due to gravity (although inertia must still be overcome). For example, to reproduce sinusoidal motion at 0.1 Hz with a magnitude of 1 m/s^2 peak requires a shaker with a displacement of at least 5 m. The same shaker will also be restricted in its capability for peak velocity (the 0.1 Hz, 1 m/s² peak signal has a peak velocity of about 1.6 m/s). An alternate approach to considering the problem is to imagine the size of a simulator required to reproduce the vertical displacement of a passenger ferry traveling in heavy seas; a situation of utmost interest to ferry operators. Investigators are restricted to performing experiments within the safe limits of their shakers, and are not necessarily able to reproduce motion that is reported in the literature from other laboratories with facilities of a different specification. The literature shows step changes in knowledge, as experiments are completed at newly commissioned facilities, often designed with specific applications in mind (e.g., military aircraft motion, tilting train motion).

3.6.1 EFFECT OF FREQUENCY OF VERTICAL OSCILLATION ON SICKNESS INCIDENCE

A comprehensive laboratory study of the relative sickness symptoms from exposure to vertical motion is that of McCauley et al. (1976). This study used 25 stimuli at frequencies from 0.083 Hz to 0.7 Hz and with peak magnitudes from 0.78 m/s² to 7.85 m/s². As would be expected, for stimuli of the same frequency, the incidence of vomiting increased with increases in vibration magnitude (Figure 3.7). At each vibration magnitude, there was a similar trend: at frequencies above 0.25 Hz, the incidence of vomiting decreased with increases in vibration frequency. Between 0.167 Hz and 0.25 Hz there was an equal sensitivity to the motion. Below 0.167

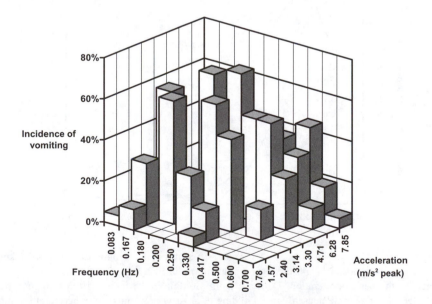

FIGURE 3.7 Incidence of vomiting associated with exposure to various magnitudes and frequencies of vertical oscillation according to McCauley et al. (1976). The frequency axis and the magnitude axis have ordinal properties only. [Adapted from Griffin, M.J. (1990). *Handbook of Human Vibration.* London: Academic Press.]

Hz, the sensitivity reduced as the frequency reduced (although this observation is only based on one data point: 0.083 Hz at 0.78 m/s² peak). These findings are in general agreement with data obtained at sea (Lawther and Griffin, 1987), although these data also had only one data point at frequencies below 0.1 Hz.

3.6.2 EFFECT OF FREQUENCY OF FORE-AND-AFT OSCILLATION ON SICKNESS INCIDENCE

For fore-and-aft oscillation, some experimental investigations have used constant acceleration across a variety of frequencies, whereas others have used constant velocity. Only one study has investigated oscillation below 0.2 Hz (Golding et al., 2001). Generally, these studies have indicated an increase in susceptibility in response to fore-and-aft oscillation at frequencies below 0.4 Hz, with the greatest susceptibility at about 0.2 Hz (Figure 3.8). The study that investigated the response at a lower frequency indicated less sensitivity at this frequency. Griffin and Mills' data (2002) show an anomaly at 0.25 Hz where the sickness incidence is less than what would be expected from the data obtained in other studies. It is possible that this difference is due to the different methodologies employed by the investigators. Golding's studies use the same 12 susceptible subjects for each frequency (i.e., a

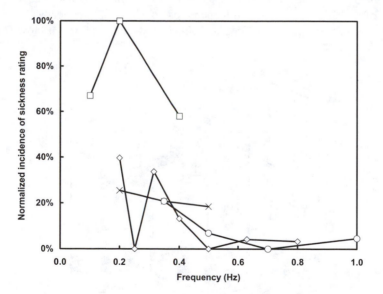

FIGURE 3.8 Effect of frequency of fore-and-aft sinusoidal oscillation on motion sickness incidence for a 30-min exposure. Results are normalized by division of the raw incidence data by the peak acceleration. Data are from Golding and Markey, 1996 (3.6 m/s² peak acceleration, moderate nausea, –×–); Golding et al., 1997 (3.6 m/s² peak acceleration, moderate nausea, –○–); Golding et al., 2001 (1 m/s² peak acceleration, moderate nausea, –□–) and from Griffin and Mills, 2002 (0.5 m/s peak velocity [data normalized according to acceleration], moderate nausea, –◇–).

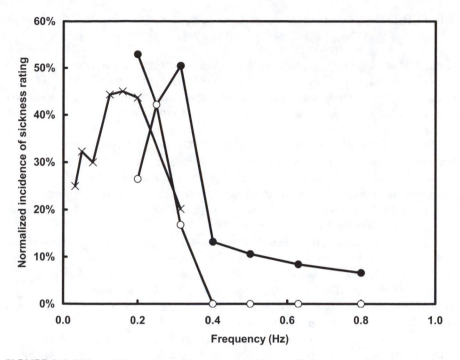

FIGURE 3.9 Effect of frequency of lateral sinusoidal oscillation on motion sickness incidence for a 30-min exposure. Results are normalized by division of the raw incidence data by the peak acceleration. Data are from Lobb, 2001 (1 m/s peak velocity at frequencies below 0.315 Hz, 0.5 m/s peak velocity at 0.315 Hz [data normalized according to acceleration], mild nausea, –×–) and from Griffin and Mills, 2002 (0.5 m/s peak velocity [data normalized according to acceleration, –●–], mild nausea,; moderate nausea, –○–).

repeated-measures design), whereas Griffin and Mills use 12 different subjects for each experimental group (i.e., an individual-measures design using a total of 192 participants) and they did not select or group subjects according to susceptibility.

3.6.3 EFFECT OF FREQUENCY OF LATERAL OSCILLATION ON SICKNESS INCIDENCE

There are only two known studies that have compared sickness incidence with the frequency of lateral oscillation (Lobb, 2001; Griffin and Mills, 2002). Both of these studies were carried out using the stimuli of constant velocity. Both the studies have shown an increased sensitivity to the motion at frequencies below 0.4 Hz with a maximum response at 0.16 Hz and 0.25 Hz (for the studies by Lobb, 2001, and Griffin and Mills, 2002, respectively; Figure 3.9). Griffin and Mills suggest that, in the 0.2 to 0.8 Hz frequency range, motion sickness incidence is independent of frequency for signals of equal peak velocity. Due to the characteristics of transforming a velocity signal to an acceleration signal, this amounts to a reduction in sensitivity with frequency for acceleration.

3.6.4 RELATIVE NAUSEOGENICITY OF LOW-FREQUENCY OSCILLATION IN DIFFERENT DIRECTIONS

Golding et al. (1995) investigated motion sickness incidence for subjects exposed to fore-and-aft and vertical-sinusoidal motion at 0.35 Hz with a magnitude of 3.6 m/s^2 peak. In addition to reporting higher incidences of sickness for fore-and-aft oscillation than for vertical oscillation, it was also shown that it took more than twice as long to reach a similar sickness rating for vertical than for fore-and-aft motion.

For horizontal motion, Griffin and Mills (2002) show no significant differences between the reported sickness incidences in the fore-and-aft and lateral directions. Similarly, Mills and Griffin (2000), failed to show a consistent trend in sickness incidence with horizontal motion direction (of three seating and vision combinations, two showed highest sickness incidences for fore-and-aft motion; one showed a higher sickness incidence for lateral motion).

Although the data in the literature comprises, in places, only a small sample of studies, most of these investigations of sickness incidence for sinusoidal stimuli in the laboratory have a common conclusion: the frequency range of greatest sensitivity to low-frequency oscillation is from 0.2 to 0.4 Hz. This fundamental finding appears to hold true, irrespective of the direction of oscillation. The laboratory studies also serve to reinforce the conclusions of Lawther and Griffin (1987), who show a remarkably similar equal-sensitivity curve through studies carried out on board a ship. Finally, the similarity of the relative nauseogenicity to motion in all directions and to the W_f frequency weighting serves to increase the confidence of those using W_f and the MSDV in applications for which they were not originally designed (i.e., exposure to vertical oscillations on board ferries) and where the dominant motion is not vertical.

3.7 HABITUATION TO MOTION SICKNESS

When exposed to a novel combination of sensory signals, the experience of the mismatched signals is memorized. Therefore, the novelty of the mismatch reduces with each repeated exposure until the sensory signals agree with the expected combination of senses. With repeated exposures, sensory rearrangement theory predicts a reduction in the symptoms of sickness.

With repeated exposures to similar forms of provocative stimuli, most people show signs of habituation within a few days. For example, most competitors in a round-the-world yacht race (see also Subsection 3.4.1 and Figure 3.5) habituated to the movement of the boat within 5 days for each of the four legs of the competition (Turner and Griffin, 1995). Both the percentage of people who show no signs of sickness and the time to report symptoms during exposure to provocative stimuli increase when subjects experience the stimuli daily (Hill and Howarth, 2000; Howarth and Hill, 1999; Figure 3.10). However, after an extended break from the nauseogenic stimulus, the habituation virtually disappears for most individuals, although habituation tends to be more rapid when reexposed (e.g., Golding and Stott, 1995). It is generally accepted that repeated exposures with a gap of more than one week are insufficient to cause habituation, although some laboratory studies have

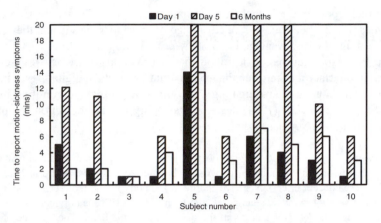

FIGURE 3.10 The time elapsed (minutes) before the first report of a motion-sickness symptom increase for 10 subjects exposed to a 20-min nauseogenic stimulus from a head-mounted display on three occasions spaced 5 d and 6 months apart [data from Howarth and Hill (1999)].

still shown some habituation with 1-week intervals for virtual stimuli (e.g., Clemes and Howarth, 2002).

3.8 PREVENTION OF MOTION SICKNESS

A motion sickness sufferer might justifiably ask, "What can I do to avoid feeling sick?" There is an enormous array of solutions available, ranging from the bizarre and those based on folklore and old wives' tales, through alternative therapies, behavioral advice, and drug treatment. Lawther and Griffin (1988) note that some ferry travelers proposed antidotes as a relief for sickness that other travelers had suggested as provocations for sickness. One strategy occasionally suggested is to eat bananas as they are reported to taste the same in both directions!

3.8.1 AVOIDANCE

The most effective way of eliminating motion sickness is to avoid situations where sickness-inducing motion might occur. Many susceptible sufferers choose to avoid certain modes of transport due to fear of sickness. As such, a degree of self-selection exists for passengers of all transport types. Where alternatives are viable, provocative situations will be avoided. For example, some individuals choose rail travel in preference to coach travel due to the perceived difference in nauseogenicity of the journeys. As a solution to sickness, avoidance might be necessary for susceptible travelers who are able to choose whether to travel or not, but this is not a practicable one for the traveler who is required to make a trip (e.g., business travel or military operations). Of course, avoidance will make habituation impossible.

3.8.2 ALTERNATIVE THERAPIES

It seems that every branch of "alternative medicine" can provide a solution to motion sickness. Some of the most common treatments are herbal (ginger and peppermint) or have their roots in traditional Chinese medicine (e.g., a P6 acupressure device, sometimes sold as "sea bands"). An advantage of these therapies is that they contain no active drug; the disadvantage is that they contain no active drug!

The P6 acupressure point lies about 5 cm proximal to the first crease at the wrist–palm interface and between the tendons that connect the digital flexor muscles to the fingers. In a study investigating the control of nausea and vomiting in pregnancy, Stainton and Neff (1994) describe the theory behind the action of applying pressure to the P6 (Neiguan) point:

> As described by Chinese traditional doctors who use acupressure and acupuncture, bilateral pressure on this point in the wrists rebalances energy (the best translation of the Chinese concept of chi). Negative energy in the body is first redirected away from the heart using the right Neiguan point, and then more positive energy comes into the body through the left Neiguan point, the one closer to the heart. The balance of the life forces yin and yang is restored, and nausea is controlled.

Although most members of the scientific community are at least skeptical of theories relating to the life forces yin and yang (this author included), the fact that there is a long history of patients reporting reduced sickness cannot be ignored. The question of why the treatment works is different from the question of whether the treatment works. There are scientific studies of the efficacy of P6 acupressure in preventing sickness of many types to be found in the related literature. For postoperative nausea, some studies have shown a benefit (e.g., Harmon et al., 2000) whereas others have not (e.g., Agarwal et al., 2000). Benefits have also been reported for reducing nausea and vomiting during pregnancy (e.g., Stainton and Neff, 1994), although other studies have shown similar benefits for a placebo group (Norheim et al., 2001).

Although advocated by many in real travel situations, alternative therapies have proved less convincing for control of motion sickness in laboratory studies, where the motion is a controlled stimulus (e.g., Bruce et al., 1990; Warwick-Evans et al., 1991). Considering that psychological factors are significant elements in motion sickness, it is possible that such therapies act as an effective placebo by giving the sufferer the confidence to travel and to partake in distracting activities. Certainly, such therapies have not been demonstrated to make the sickness worse. Considering that there are no side effects, and also that the possibility of benefits cannot be ruled out, a susceptible traveler could try such an alternative therapy to investigate whether any relief is felt, even if it is due to a placebo effect.

3.8.3 ANTIMOTION-SICKNESS DRUGS

A range of over-the-counter and prescription drugs are available for the motion sickness sufferer. The drugs work as either antihistamines or anticholinergics, or have elements of both actions. The mechanism by which antihistamines are effective in preventing motion sickness is unclear, but other forms of antihistamines are effective in reducing allergic reactions to toxins such as insect stings. Anticholinergic reactions reduce muscle activity in the stomach.

The most effective drug for prevention of motion sickness is scopolamine (also known as hyoscine or "Kwells"). This is an anticholinergic treatment which is effective for short periods of time following administration by either tablet or injection. It is also possible to obtain the drug in a small transdermal patch which is worn behind the ear (e.g., Parrott, 1989). This is known as "transderm scop" and can provide protection for up to 72 h by supplying a continuous controlled dose of the drug. Side effects can include dryness of the mouth, drowsiness, and blurred vision (i.e., some of the symptoms that occur if affected by motion sickness!). Therefore, those using scopolamine should not drive or carry out safety-critical tasks. Scopolamine is usually only available on prescription.

Common antihistamine drugs for prevention of motion sickness include cinnarizine (Stugeron), meclozine (Sea-legs), dimenhydrinate (Dramamine), and promethazine (Phenergan). Although not usually as effective as scopolamine, antihistamines have fewer side effects (usually only drowsiness), some have longer lasting effects, and some formulations have, in certain countries, been approved for use with children above the age of 2 years.

One side effect that might not be immediately apparent is that if antimotion sickness drugs are effective, then habituation is slower or even nonexistent. Therefore, in the long term, drugs might not be as effective a treatment as simply suffering the symptoms and allowing habituation to occur with repeated exposure.

In a study of pharmaceutical use by U.S. astronauts on space shuttle missions, the most common type of drug taken was to counter space motion sickness (Putcha et al., 1999). Of the 219 person-missions studied, antimotion sickness drugs were reported as being taken in 44% of the medical records. The preferred drug was promethazine taken orally, although it was also administered intramuscularly and rectally. Drug doses were considered effective in more than 85% of incidents studied, the best results being reported for intramuscular administration. Two thirds of the doses were taken on the day one of the space mission with a rapid reduction in dosage for subsequent days. This could be due to habituation to the sensory conflict, although it is possible that astronauts were exposed to more nauseogenic stimuli during launch or that they took extra precautions during their first mission day. However, patterns of drug usage for other types of medication during the missions did not follow a similar pattern.

All motion sickness drugs must be administered at least 1 h before travel. They are not effective once any symptoms have occurred. As such, they are effective as a preventative therapy only, and users must anticipate the exposure to provocative motion. As for all types of pharmaceutical interventions, motion sickness drugs do not affect all people in the same way; some people experience a great relief from the sickness whereas others feel little effect.

If a traveler is susceptible to motion sickness then preventative drugs are more likely to be taken. Here lies another motion sickness paradox: in studies of motion sickness incidence among travelers, there is a positive correlation between sickness incidence and the use of antimotion sickness drugs (e.g., Turner and Griffin, 1995; Turner and Griffin, 1999). This does not indicate a failing in the efficacy of the drugs, but successful self selection of the susceptible. Of course, it is also possible that these susceptible travelers do not all consume the drugs in time for them to become effective, therefore limiting their ability to protect.

3.8.4 Optimizing the Motion Stimulus

As for all other aspects of vibration exposure, there are two strategies which could be employed to optimize the stimulus for human exposure. The first is to reduce the magnitude of the motion; the second is to design or modify the environment or vehicle such that the motion occurs at frequencies where the body is less responsive.

For land transportation, a reduction in acceleration magnitudes is possible by reducing speed while cornering and by reducing the acceleration and braking forces (see Subsection 3.4.3). Changing the dominant frequency of vibration can be achieved by vehicle design or route design. For example, by investigating the distance between successive bends, it is possible to calculate the dominant frequency of lateral acceleration for railway motion as the designer will know the speed of the trains which will use the track. This approach can also be applied to roads (e.g., Losa and Ristori, 2002). As a general principle, motion at frequencies close to 0.2 Hz should be avoided.

For air transportation, low-frequency motion occurs while changing direction or when flying through turbulence. Repeated direction changes can be extremely nauseogenic for small military jet aircraft, but in these cases, the direction of the aircraft is under the control of a pilot who is in constant communication with the crew and so there is scope for altering the flight path should any member of the crew experience symptoms of sickness. For civilian aircraft, there are fewer direction changes, but the pilot is removed from the immediate issue of discomfort of the passengers. Furthermore, the route is also partially prescribed by air traffic control and so there is less scope for altering the motion stimulus should any occupant feel nauseous. However, unless a long period of waiting for a landing slot is forced upon the aircraft, civilian air travel does not involve a large number of direction changes. Most nausea for passenger aircraft is caused by the low-frequency vertical motion induced by flying through air turbulence (Griffin, 1990). There is little that can be done once turbulence has been encountered; regions of atmospheric instability should be avoided by changing route or altitude.

3.8.5 Avoiding Conflict

One effective method of reducing motion sickness in many environments is by taking steps to avoid sensory conflict. For most types of travel sickness, the conflict is visual–vestibular Type I or Type IIb. Therefore, if a reliable visual scene which gives appropriate motion cues can be viewed, then the conflict will be reduced or eliminated. Conversely, if the attention is focused on a visual scene which provides no motion cues, then the strength of the conflict will be enhanced. For example, reading, writing, and using portable computers or handheld electronic games can increase sickness in many forms of passenger transport. These activities are often employed as means of reducing the tedium of the journey or as a means of optimizing the time spent traveling. It could be argued that, in some forms of transport (e.g., boats or aircraft), performing a distracting activity (e.g., working or reading) produces no more sensory conflict than would be experienced otherwise, due to the individual traveling inside the vehicle and that the interior of the cabin does not provide any visual cues to reinforce the vestibular and somatic sensation of the motion. If there is no vestibular input (i.e., no acceleration, although a constant velocity might exist), then, according to sensory conflict theory, there is no reason why an activity cannot be performed with a fixed visual scene. However, if the traveler is concerned about minimizing chances of feeling motion sickness, then it is wise to try to provide a reliable visual cue, should acceleration occur (e.g., looking out of the window at landscape objects).

Babies are usually unsusceptible to motion sickness but children from the age of about 2 years can experience symptoms (Griffin, 1990). This is probably due to the time taken to establish an expected pattern of sensory inputs (Brandt et al., 1976). Preventative action can be taken to help reduce the sensory conflict of children by giving a clear external view (e.g., remove sunshades in cloudy weather, provide an elevated child seat with a good view through the windows), by encouraging focusing on stationary external objects (e.g., the passing scenery), and by discouraging the use of handheld games and reading books or magazines while traveling. Playing "I-spy" type games or listening to prerecorded audio books or music can be suitable alternate activities for some children. As discussed in Subsection 3.4.3, the driver is at least partly responsible for the low-frequency acceleration stimulus, and so it is possible for the driver to help minimize the vestibular component contributing to the sensory conflict for the children in a vehicle. This can be achieved by driving smoothly, especially on winding roads. Although young children commonly fall asleep while traveling, the other symptoms of sickness (see Section 3.2) should be monitored throughout the journey (especially for the susceptible), so that if symptoms develop, action can be taken. This could mean taking a break or, at worst, being prepared to ensure that the ultimate unpleasant endpoint is not reached within the close vicinity of other travelers (or their expensive car interiors)!

3.8.6 MOTION SICKNESS DESENSITIZATION PROGRAMS

Approximately 15% of trainee U.K. Royal Air Force aircrew do not adapt to the motion of their aircraft (Carr, 2001). Therefore, airsickness can become a problem. Although drug treatment can help during training, it is not an acceptable long-term solution due to the side effects. One option is to retrain the crew to perform ground duties only; the other option is to enroll the crew in a desensitization program to enforce habituation to the provocative motion. The U.K. Royal Air Force program (which has been running since 1966) consists of two phases: a ground phase (for all aircrew) and an air phase (for fast-jet pilots and navigators only). During the ground phase, aircrew are exposed to nauseogenic stimuli in a laboratory twice a day for 4 weeks to gradually increase their tolerance to the motion (Bagshaw and Stott, 1985; Carr, 2001). The simulated motion is designed to represent the type of motion in the aircraft that causes them sickness. The air phase consists of a further 4-week program of training in a fast jet including flying high-speed, banked turns, instrument flying (zero visibility), and aerobatics. In both phases, the provocative stimulus is terminated when participants feel moderately nauseous (i.e., prior to vomiting). This desensitization strategy has proved very effective with over 90% of those treated losing their susceptibility to airsickness.

Although access to such programs is only available to a lucky (or unlucky!) few, some of the principles can be applied generally. For example, if an individual changes employment to work in a potentially nauseogenic environment, then they might take some consolation in appreciating that repeated exposure to the motion almost always results in desensitization, even for the most susceptible.

3.9 CHAPTER SUMMARY

Motion sickness is a normal response to a novel combination of sensory signals. A wide range of symptoms can develop, the most obvious of which is vomiting. Sensory rearrangement theory states than when a battery of sensory signals are received by the brain which are in conflict with the expected combination of sensory signals, then nausea might result. It is hypothesized that this response is due to the sensory conflict being misinterpreted as a result of toxin ingestion.

Sickness can occur in most forms of passenger transport including sea, land, air, and space travel. Studies at sea have shown that humans have a peak in sensitivity at about 0.2 Hz. This finding has been reinforced by laboratory studies. The BS6841 W_f frequency weighting and the motion sickness dose value can be used to predict incidence of sickness for ferry passengers. On the road, one important parameter in the nausogenicity of any journey is the driving style. Therefore, the driver is able to minimize the chances of passengers developing motion sickness symptoms. With frequently repeated exposures to motion, most people habituate to the sensory conflict, but the habituation can wear off if the motion stimuli are not experienced regularly. It is possible to reduce the cause or effects of motion sickness by the design of vehicles or routes, behavioral factors, and by the use of drugs. Although some individuals are helped by the use of alternative therapies, it is likely that the improvement is a placebo effect.

4 Hand-Transmitted Vibration

4.1 INTRODUCTION

Hand-transmitted or hand-arm vibration occurs whenever an individual holds a vibrating tool. This could be a powered surgical instrument, floor polisher, demolition pick, pavement breaker, riveting gun, motorcycle handlebars, hair clipper, or a wide variety of other types of tools in a broad cross section of industries.

Hand-transmitted vibration is an industrial phenomenon. Pneumatic tools are first reported being used in French mines in 1839, and the use of vibrating tools gradually spread to other industries. Miners, shipbuilders, and chainsaw operators have been traditionally associated with vibration-induced health problems, although it is now accepted that many more sectors are affected. Improvements in understanding the health risks and the increased availability of low-vibration emission tools can give us some optimism that the incidence of hand-transmitted vibration injuries should gradually reduce. With the trend towards globalization it is likely that populations with vibration problems will become more apparent in the developing world due to the currently less sophisticated health and safety cultures in these nations.

The two terms *hand-arm vibration* and *hand-transmitted vibration* are synonymous. Hand-transmitted vibration is the term preferred by the author, as this clearly indicates the source of the vibration. Furthermore, almost all of the health effects of the phenomenon are localized in the hand. Nevertheless, there is currently no consensus, and both terms are commonly used in the literature.

Someone using a vibrating tool perceives the vibration through tactile receptors in the skin (Section 4.2), but there is a risk of damage to both the vascular and neurological systems (Section 4.3). To minimize the risk, a range of actions should be implemented in parallel (Section 4.5). If, despite these actions, a problem becomes apparent, then medical professionals might be required to diagnose and classify the nature and extent of the problem (Section 4.4).

4.2 HAND-TRANSMITTED VIBRATION PERCEPTION

The sense of touch is perceived through a combination of many sensory pathways. Sensory signals through our hands are interpreted as texture, shape, temperature, location, size, movement, and, in some situations, pain. For tactile perception in glabrous skin (i.e., hairless skin such as the palms of the hands), receptors are

TABLE 4.1
Types of Tactile Receptors Found in the Hand

Adaptation Speed	Psychophysical Channel	Receptor Ending	Receptive Field Size	Frequency Range (Hz)
Fast acting I	Nonpacinian I	Meissner	Small	5–60
Fast acting II	Pacinian	Pacinian	Large	40–400
Slow acting I	Nonpacinian III	Merkel	Small	0–5
Slow acting II	Nonpacinian II	Ruffini	Large	100–500 (also pressure and stretching)

embedded within the skin and are referred to by receptor ending, speed of action, or psychophysical channel (Table 4.1). The exact frequency range of sensitivity of each channel is not well defined, partly due to the difficulty in stimulating any one channel in isolation and also due to interindividual differences. Type I receptors are found close to the skin surface and therefore are able to sense location of sensation more precisely than Type II (Figure 4.1). Type II receptors are found deeper in the dermis, and therefore sensation is more general in terms of receptive field size.

Most of the literature relating to the perception of hand-transmitted vibration has been based on psychophysical experiments in which subjects are exposed to vibration with an extended hand resting lightly on a vibrating probe. The detail of the experimental design can be critical in obtaining results repeatable between studies (Lindsell and Griffin, 1998). Although the general shape of the perception threshold or equal sensation curves remain generally similar, increasing the area of contact can increase the vibration sensation (e.g., Harada and Griffin, 1991; Maeda and Griffin, 1994; Whitehouse, 2002). Similarly, probe design and push force on the probe can affect

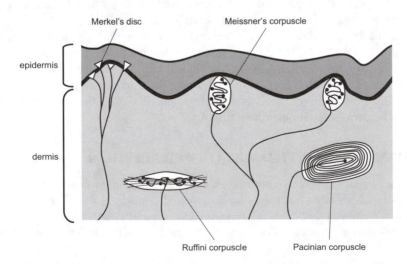

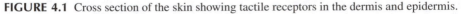

FIGURE 4.1 Cross section of the skin showing tactile receptors in the dermis and epidermis.

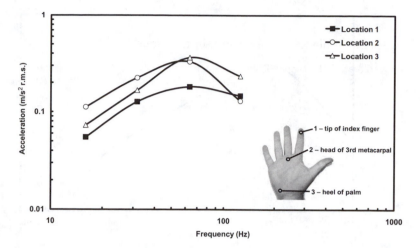

FIGURE 4.2 Mean vibration perception thresholds as a function of frequency for three measurement locations on the hand. Data from Morioka (1999).

thresholds. Finally, the precise location on the hand of the test can also affect results (e.g., Morioka, 1999; Morioka, 2001; Figure 4.2). So, if we are interested in testing whether or not an individual can perceive the vibration to which they are exposed, we must consider not only the magnitude of the vibration but also the frequency of the vibration and nature of the contact, in addition to interindividual differences.

A single frequency weighting, termed W_h, is used for hand-transmitted vibration, irrespective of its direction (for a critique of the applicability of frequency weightings in general, see Section 2.2.4). The weighting is designed to simulate the response of the hand-arm system to vibration and can be applied when assessing the risk of injury from vibration or when an indication of the sensation magnitude is required. The weighting is defined in ISO 8041 (1990) and ISO 5349-1 (2001). Between 8 and 16 Hz the idealized weighting is "flat" (i.e., the human response to vibration is similar at any frequency within this range). Between 16 and 1000 Hz, the idealized weighting has a roll-off rate of 6 dB/octave (i.e., as the frequency increases, the human response to vibration decreases). Below 8 Hz and above 1000 Hz, ISO 5349-1 states that "the frequency dependence is not agreed," and therefore vibration outside this frequency range should not be included in analyses based on standardized techniques.

In practice, frequency weightings are usually implemented using digital or analog signal-processing techniques. A "realizable" W_h weighting consisting of a series of curves defined by characteristics of a filter has also been defined. The idealized and realizable versions of W_h frequency weighting are illustrated in Figure 4.3.

4.3 HEALTH EFFECTS OF HAND-TRANSMITTED VIBRATION

Most of the interest with hand-transmitted vibration is due to the disorders that are often observed in populations who use vibrating tools. Disorders can be broadly

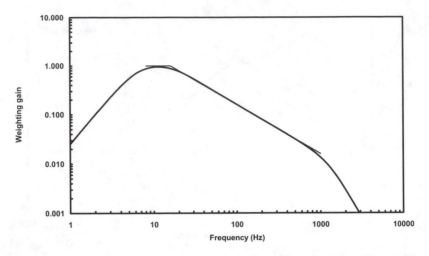

FIGURE 4.3 Idealized (straight line) and realizable (curved line) W_h frequency weighting as defined in ISO 8041 (1990).

divided into vascular and nonvascular categories (Griffin, 1990). Nonvascular disorders can be further subdivided into bone and joint, neurological, muscular, and other disorders. Collectively, both types of symptoms are referred to as *hand-arm vibration syndrome* or *HAVS*.

4.3.1 VASCULAR ASPECTS OF HAND-ARM VIBRATION SYNDROME

The most well-known clinical disorder caused by vibration exposure is *vibration-induced white finger* or *VWF*. Sometimes, this is known as "dead man's hand" or Raynaud's disease of occupational origin.

4.3.1.1 Primary Raynaud's Disease

Maurice Raynaud describes discoloration of the digits (blanching or whitening) in his 1862 doctoral thesis, *De l'asphyxia locale et de la gangrene symetrique des extremites*. The name Raynaud's disease was used to conveniently label a variety of conditions with unknown origin in which intermittent pallor occurred in the nose, ears, hands, and feet. As understanding of the causes of blanching increased, the number of cases of "unknown origin" reduced, and so primary Raynaud's disease has now become synonymous with *constitutional white finger*, that is, a condition with no pathological cause but caused by genetic factors (which is often observed occurring within family clusters).

Primary Raynaud's disease is characterized by intermittent bilateral blanching of the fingers. During recovery, the fingers often turn red, and this can be the most painful component of an attack. Most sufferers are only affected in the hands, although some report blanching of the toes, nose, ears, or lips. Attacks are triggered by exposure to cold. The prevalence of the disease is about 10% in females and 1 to 5% in males, although these figures are still open to some debate.

For Raynaud's disease to be "primary," it should occur bilaterally and should have occurred since young adulthood, eliminating other possible causes.

4.3.1.2 Secondary Raynaud's Disease

Secondary Raynaud's disease is, by definition, one in which the symptoms have a known cause. The most common cause is vibration exposure, although there are a variety of other possibilities (Table 4.2). These can be broadly categorized into trauma, intoxication, obstructive arterial diseases, connective tissue diseases, and a hypersensitivity in response to cold (Pelmear and Wasserman, 1998).

4.3.1.3 Vibration-Induced White Finger (VWF)

Vibration-induced white finger is a form of secondary Raynaud's disease. The first clear vascular symptom of VWF is usually an intermittent blanching of a fingertip, provoked by cold. With continued exposure to vibration, the area affected by blanching increases, spreading to other fingers, affecting more of each digit, and ultimately affecting all phalanges from fingertip to palm. One distinguishing feature of the blanching is that there is a clear boundary between the blanched and normal parts of the finger. The severity of the condition can therefore be assessed by counting the number of fingers affected and examining whether the proximal, medial, or distal phalanges blanch. Attacks generally last less than 30 min, and the most painful element of an attack is often when the blood returns to the fingers as the blood vessels dilate. This can be combined with the fingers turning red as they are refilled with blood. In the worst cases, necrosis can occur, whereby the tips of the fingers are irreversibly damaged, and the tissue dies.

It is important to note that the damage is done by vibration, but the attacks are triggered by cold. Therefore, a sufferer might not see the symptoms in a warm work-environment and may be unable to exhibit blanching to a company health professional or general practitioner during a consultation. It is often leisure activities that are most affected, with the sufferer avoiding pursuits that might involve exposure to cold (e.g., skiing, fishing, gardening, outdoor sports).

It is a normal homeostatic response of the body to induce peripheral vasoconstriction during exposure to cold so that core temperature can be maintained. For an individual with VWF, the vasoconstriction is exaggerated to the point of occlusion. The physiological mechanism underlying this response is not entirely clear, but it is likely to be due to either damaged capillaries or an over-response of the sympathetic nervous system (Pelmear and Wasserman, 1998).

Usually, there is a latent period of some years between the first exposure to vibration and symptoms, although this period can be as short as a few months, and other workers appear resistant to VWF. For example, for caulkers and riveters working in a dockyard, 20% of workers reported VWF after 6 years of exposure, rising to 50% after 18 years (Nelson, 1988).

Models of exposure–response relationships to link vibration with prevalence of VWF have been suggested (e.g., Brammer, 1986). These suggest increased risk with increased vibration magnitude and exposure time. It remains unclear whether the

TABLE 4.2
Summary of Causes of Raynaud's Disease

1. Primary Raynaud's disease
    a.   Constitutional white finger
          i.   Genetic propensity to Raynaud's
2. Trauma
    a.   Direct to extremities
          i.   Hand-arm vibration syndrome (HAVS)
          ii. Lacerations with blood vessel injury
          iii. Frostbite and immersion syndrome
    b.   Compression of proximal blood vessels
          i.   Thoracic outlet syndrome
          ii. Cervical rib
          iii. Costoclavicular and hyperabduction syndrome
3. Intoxication
    a.   Drug medication
          i.   Beta-adrenoreceptor-blocking agents
          ii. Ergot preparations
          iii. Methysergide
          iv. Bleomycin and Vinblastine
    b.   Vinyl chloride
4. Obstructive Arterial Diseases
    a.   Arteriosclerosis obliterans
    b.   Thromboangiitis obliterans
    c.   Arterial emboli
5. Connective Tissue Diseases
    a.   Rheumatoid arthritis and Sjögren's syndrome
    b.   Scleroderma
    c.   Systemic lupus erythematosus
    d.   Mixed connective tissue disease
    e.   Polymyositis and dermatomyositis
6. Hypersensitivity
    a.   Cryoglobulinemia
    b.   Cold agglutinins
    c.   Cryofibrinogenemia
    d.   Paraproteinemia

Source: Adapted from Pelmear, P.L. and Wasserman, D.E.
(1998). *Hand-Arm Vibration: a Comprehensive Guide for
Occupational Health Professionals.* (2nd ed.). Beverly Farms,
MA: OEM Press.

W_h frequency weighting provides a better measure of risk than using unweighted
vibration and whether data obtained for workers exposed to very high vibration
magnitudes or shock-type vibration can be extrapolated to more moderate and
continuous exposures. Notwithstanding these difficulties, a version of the model
published in ISO 5349 (1986) provides risk estimates that have been shown to either

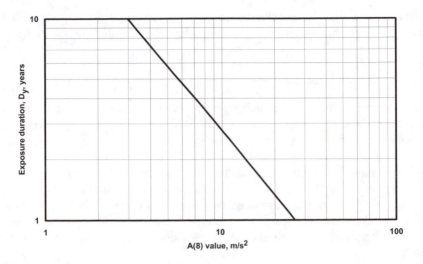

FIGURE 4.4 Vibration magnitudes and years of exposure for a 10% prevalence of vibration-induced white finger in a group of exposed persons, as predicted by the model specified in ISO 5349-1 (2001).

under- or overestimate risk (Bovenzi, 1998). One might therefore consider the ISO model (and its updated version published in 2001) as being an appropriate approximation of the risk for a population, although it must be stressed that the risk estimates cannot be used to assess the risk to an individual worker. For multiaxis frequency-weighted vibration measures, the daily vibration exposure, $A(8)$, estimated to produce VWF in 10% of the exposed population can be expressed as (ISO 5349-1, 2001):

$$D_y = 31.8[A(8)]^{-1.06}$$

where D_y is the lifetime exposure in years, and $A(8)$ is the daily vibration exposure (Figure 4.4).

So long as the condition has not developed to its most severe stages, cessation of vibration exposure can lead to recovery in some sufferers. For example, in a follow-up study of chainsaw operators, Futatsuka and Fukuda (2000) show that 25 years after cessation of vibration exposure, 40% of those with early stages of VWF and 13% of those with advanced VWF had recovered, with 50% of the recovery occurring within 5 years of the cessation of vibration exposure. Unfortunately, this and other studies still report that the majority of those affected do not show signs of recovery.

4.3.2 Nonvascular Disorders of Hand-Arm Vibration Syndrome

In addition to VWF, a variety of other disorders are common elements of HAVS. Some of these are directly related to vibration; others are observed among users of handheld tools that do not vibrate. The disorders can be broadly grouped into neurological, bone and joint, muscular, and other disorders that are remote from the hand and arm.

4.3.2.1 Neurological Disorders

Those exposed to hand-transmitted vibration often report numbness and tingling. Frequently, these are the first symptoms of HAVS and can be experienced before vascular disorders. It is common for these symptoms to occur after using vibrating tools, even in a novice operator or domestic DIY user of powered household or gardening tools. Therefore, it is difficult to use numbness and tingling as an indicator of susceptibility to HAVS, *per se*. However, if the symptoms occur periodically without being immediately preceded by vibration exposure, then the symptoms should be taken more seriously. More severe damage to the peripheral nerves can result in intermittent or persistent reduced sensory perception, reduced tactile discrimination, and impaired manual dexterity.

Another type of neurological disorder that is often observed with those who are exposed to vibration is carpal tunnel syndrome (CTS). The carpal tunnel is formed between the bones of the wrist and the transverse carpal ligament through which the flexor tendons and the median nerve pass. If the median nerve is injured, then numbness and loss of feeling and grip can occur. CTS is also common among those who operate all types of tools, and so it is possible that the condition is a hazard of the job in general, rather than a symptom due to vibration itself. This is one example of why it is important to consider all ergonomic risk factors when evaluating a task and not to concentrate on just one aspect of the total problem.

4.3.2.2 Bone and Joint Disorders

Griffin (1990) lists more than 100 reports (published between 1926 and 1987) of bone and joint disorders among workers exposed to vibrating tools. One common disorder was damage to the lunate bone of the wrist (Kienböck's disease). This condition is thought to be partly caused by a reduced blood supply to the bone in addition to compression micro fractures. Other common bone disorders include cysts, although these are common among users of nonvibrating tools, too.

Some investigators in the industrial health discipline have reported osteoarthritis or decalcification of bone in those individuals exposed to vibration. It is interesting to observe that those researching in other disciplines (e.g., bone growth and remodeling) have observed an increase in bone density with vibration exposure of the limb (e.g., Rubin et al., 2004; Verschueren et al., 2004).

4.3.2.3 Muscular Disorders

Loss of grip strength is a very common impairment in those exposed to vibration (e.g., Färkkilä et al., 1982). Loss of strength can initially occur immediately after exposure to vibration but can become more persistent with continued exposure. It is suggested that the cause of the reduced grip force is due to the incomplete contraction of muscle.

Occasionally, tendons or their synovial sheaths can become inflamed causing tendonitis or tenosynovitis, respectively. Conditions such as lateral epicondylitis ("tennis elbow") or medial epicondylitis ("golfer's elbow") are common forms of tendonitis observed with users of tools. Similarly, tool users commonly exhibit painful swelling of the extensor tendon sheaths of the thumb (de Quervain's disease) or the flexor tendons of the finger ("trigger finger"). As for other elements of HAVS, these conditions are unlikely to be directly caused by vibration but by other ergonomic factors (e.g., posture, manual handling) associated with the job.

4.3.2.4 Disorders Remote from the Hand and Arm

A range of other disorders have been associated with hand-transmitted vibration, which do not affect the hand or arm. One could argue that these should not be included in the symptomatology of HAVS as their link with vibration is often tenuous, and hand-transmitted vibration energy did not precipitate the damage reported.

Many vibratory tool users also show signs of hearing loss. This is not surprising, considering that many vibrating tools are also very noisy! However, some investigators have suggested that the addition of hand vibration to a noise increases the risk of hearing damage (e.g., Pyykkö et al., 1982; Palmer et al., 2002). It is hypothesized that a noise-induced sympathetic response of the cochlea to the vasoconstriction in the hands increases the risk of hearing damage.

Prevalence of low back pain is often higher among those exposed to hand-transmitted vibration. It is possible that the vibration is transmitted through the arm and to the lumbar spine to directly cause the injury. However, it is perhaps more likely to be associated with maneuvering heavy tools or the poor postures required to complete the tasks. For example, modern hydraulic breakers can weigh well over 30 kg.

Abdominal injuries have been reported from workers leaning on tools. For example, in a letter to the *New England Journal of Medicine*, Kron and Ellner (1988) describe a condition termed "buffer's belly" observed in a patient who leaned on a floor polishing machine. Other machine operators are known to increase the push force by pressing the torso against the tool or hands. This technique cannot be recommended as this will increase the transmission of vibration to the internal organs.

Some individuals required to stand on vibrating surfaces have developed *vibration-induced white toe* in which the toes blanch in a similar way to the fingers of a VWF sufferer during an attack. Surfaces can vibrate due to them being rigidly coupled to a mounting point of a heavy tool (e.g., a platform mounted mining drill; Hedlund, 1989), or it could be a foot pedal that vibrates (e.g., Tingsgård and Rasmussen, 1994).

4.4 CLASSIFICATION AND DIAGNOSIS OF HAND-ARM VIBRATION SYNDROME

Some industrial diseases and injuries are straightforward to classify and diagnose. For example, accidental amputation of a fingertip cannot be disputed. HAVS is far more difficult to classify, partly due to the effects usually being episodic. It is often not possible for a medical professional to observe an attack of VWF and, without specialist equipment, it is not possible to make objective measurements of sensory function. An affected individual might be well advised to try to obtain photographs of the extent of blanching as this can be helpful as evidence of the extent of their affliction. It is not so easy to provide evidence of the reduction in sensory function.

Methods for assessing the extent of HAVS can be categorized into checklist-type methods and objective measurement methods. Checklist methods rely on the honesty of the patient (which is not always guaranteed if there is a potential compensation claim) but are easy to complete. Objective measurements are claimed to be more reliable but usually require expensive sensory testing equipment and a trained operator. Neither of these methods are infallible, and therefore they should be used as indicators only and not as pass or fail criteria.

4.4.1 CHECKLIST METHODS OF CLASSIFYING HAND-ARM VIBRATION SYNDROME

The most common technique for classification of the VWF is to use the Stockholm Workshop scale, which was proposed and accepted at a meeting held in Sweden in 1986 (Gemne et al., 1987, Table 4.3). The classification is based on a medical history given by the affected individuals and results in grading according to one of the five points in the scale. Stage 0 corresponds to no attacks, and the grading increases with the severity and frequency of attacks. Stage 4 is straightforward to diagnose as trophic changes in the fingertips are usually easy to identify, although very rare. The choice of the other possible stages of the Stockholm scale is a matter of judgment for the assessor. The staging is completed for each hand separately, and the number of fingers affected is counted. A classification of 1L(2)/2R(2) means that the left hand is at Stage 1 for two fingers and that the right hand is at Stage 2 for two fingers.

One particular problem with the scale is the simultaneous assessment of frequency of attacks and the extent of the affected area. It is therefore possible for an individual to alter their staging simply by reducing (or otherwise) the frequency of their exposure to cold, thereby reducing (or otherwise) the frequency of potential attacks. Also, a well-informed worker could artificially report more severe or more frequent attacks in the hope of receiving a larger compensation payout.

TABLE 4.3
The Stockholm Workshop Scale for the Classification of VWF

Stage	Grade	Description
0		No attacks
1	Mild	Occasional attacks affecting only the tips of one or more fingers
2	Moderate	Occasional attacks affecting distal and middle (rarely also proximal) phalanges of one or more fingers
3	Severe	Frequent attacks affecting all phalanges of most fingers
4	Very severe	As in Stage 3, with trophic skin changes in the fingertips

TABLE 4.4
Brammer et al.'s Scale for the Classification of Sensorineural Stages of HAVS

Stage	Description
0SN	Exposed to vibration but no symptoms
1SN	Intermittent numbness, with or without tingling
2SN	Intermittent or persistent numbness, reduced sensory perception
3SN	Intermittent or persistent numbness, reduced tactile discrimination or manipulative dexterity or both

At the same 1986 Stockholm Workshop where the vascular scale was proposed, Brammer et al. (1987) suggested a similar type of scale for application to sensorineural aspects of HAVS. The stages progress from 0SN, where the individual is exposed to vibration but has no symptoms, through to 3SN for the most severe cases (Table 4.4). It is possible that at stage 1SN, a patient presenting to a medical professional might have normal responses to basic sensory tests, due to the intermittent nature of the condition. The staging is again completed for each hand separately.

Although staging of HAVS provides an excellent tool for simple classification of the severity of the disease, it is difficult to closely monitor an individual's deterioration or improvement, using the technique in isolation. Staging also places a heavy reliance on interpretation of the descriptors and does not provide a system of notation to identify exactly where blanching (or sensory impairment) occurs.

Griffin developed a method of scoring that can be used to map where symptoms occur (Griffin, 1990). Although this system is usually used for blanching, it can also be applied to any type of symptom that occurs to the fingers. The system is illustrated in Figure 4.5 where the shaded areas correspond to those regions where blanching

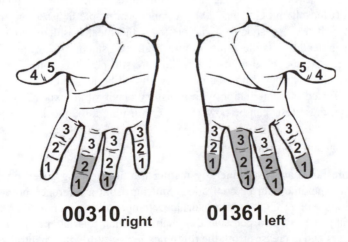

FIGURE 4.5 Griffin's method of scoring the areas of the digits affected by blanching. Shaded areas correspond to those regions where blanching occurs.

occurs. Each hand is scored with a sequence of five numbers corresponding to the thumb and each individual finger. For each finger, a score of 1, 2, or 3 corresponds to blanching occurring on the distal, medial, or proximal phalanx, respectively. For the thumb, the scores are 4 and 5. For each digit, the scores are summed to provide a single numeric indicator for the nature of the blanching. If it is assumed that blanching does not occur solely in the proximal phalanx of any finger, then the scores can be decoded to produce a complete picture of the blanching pattern. It was originally suggested that the scores across each hand could be summed to provide a total score out of a maximum of 33 for the left and right sides. Unless the complete numerical sequence is retained, this simplification process is unwise because such data compression deletes the important information that allows for mapping of the patterns of blanching, which is the distinctive quality of Griffin's method.

4.4.2 Objective Methods for Diagnosing Hand-Arm Vibration Syndrome

There are a range of available objective techniques and tests for assessing vascular and neurological components of HAVS. Combining a selection of the most effective tests into a single test-battery improves the reliability of the diagnosis, although each test has individual value. A standardized test-battery combination includes neurological tests (thermal thresholds and vibrotactile thresholds) in addition to vascular tests (rewarming times and finger systolic blood pressures) (Lindsell and Griffin, 1998). Care must be taken in all such physiological tests to avoid false positive or false negative results. Therefore, subjects should be allowed to adapt to the room conditions for 15 min prior to testing. Room temperature should be 20 to 24°C, and there should be no drafts. Background noise should be constant at about 50 dB(A) with no sudden sounds. Stimulants or sedatives should be avoided prior to testing (e.g., caffeine, tobacco, alcohol). It is recommended that neurological tests are completed prior to vascular tests, as the vascular tests might influence the neurological results (Lindsell and Griffin, 2000). A general principle in assessing the results is that more than one standard deviation from the mean result in an unaffected population indicates possible injury; more than two standard deviations indicate definite injury. Although false positive results are still possible using these criteria, among normally distributed populations they should be less than 3% for each test. Therefore, if used in combination with a history of vibration exposure and reports of blanching, these tests can be powerful in confirming a diagnosis.

4.4.2.1 Thermal Threshold Testing

Thermal thresholds are determined by resting the finger being tested on a surface where the temperature can be controlled. Starting from a reference point (usually skin temperature or 32°C) the contact surface's temperature is increased or decreased until the subject is able to perceive the change. When this has been repeated for both increases and decreases from the reference temperature, the "neutral zone" can be calculated in which the subject is unable to perceive thermal changes.

A normal neutral zone is about 15°C. If the zone exceeds 21°C, then it is possible that damage has occurred; if the zone exceeds 27°C, then damage is indicated.

4.4.2.2 Vibrotactile Threshold Testing

A tactile vibrometer is a device that tests the thresholds of vibration sensation at the fingertip. It consists of a vibrating probe on which a subject rests the fingers, or pushes with a predetermined force. The probe is made to vibrate, and the subject indicates when the vibration can be felt. Although some equipment can be configured to test at any arbitrary frequency of vibration, standard tests are carried out at 4 Hz, 31.5 Hz, and 125 Hz. These frequencies are selected to test the function of Merkel's disks, Meissner's corpuscles, and Pacinian corpuscles, respectively (ISO/FDIS 13091-1, 2001).

At each test frequency, the intensity of vibration of the probe is increased from a level below the threshold. When the vibration is perceived, the subject is required to push a response button, after which the vibration magnitude is reduced. When the vibration can no longer be felt, the subject releases the response button, and the vibration magnitude is increased again. This sequence is repeated until a repeatable set of measures has been obtained. The threshold is the mean of the levels at which the subject responded (both "can feel" and "cannot feel" levels). This is known as the von Békésy method or the up-and-down method of limits.

The mean vibration perception threshold at 4 Hz is about 0.01 m/s² r.m.s.; at 31.5 Hz, it is about 0.1 m/s² r.m.s.; and at 125 Hz, it is about 0.3 m/s² r.m.s. (ISO/FDIS 13091-2, 2001). Damage is possible if the thresholds are above 0.3 and 0.7 m/s² r.m.s. at 31.5 and 125 Hz, respectively. Damage is definite if the thresholds are above 0.4 and 1 m/s² r.m.s. at 31.5 and 125 Hz, respectively (Lindsell and Griffin, 1998).

4.4.2.3 Rewarming Time Testing

Rewarming time or "cold provocation" testing assesses the vascular component of HAVS. One or both hands are cooled, and the time taken for each finger to return to its neutral temperature is measured. Details of test methods are quite different between research groups (Harada, 2002). Some have tested in water at 0°C, which causes pain, especially for those with vascular dysfunction. Some trade-off must be made between improving the reliability of the test and maintaining its acceptability in a clinical setting. A reasonable compromise is to standardize the tests to cooling hands by immersing in water bath maintained at 12°C (ISO/DIS 14835-1, 2004). Thermocouples or thermistors are attached to the fingers, and the hand kept dry by wearing of a loose-fitting, thin, waterproof glove. Temperatures are usually measured for the duration of a settling, immersion, and recovery period (e.g., 2, 5, and 15 min, respectively).

Typical temperature profiles for cooling and recovery are shown in Figure 4.6. Temperatures for a normal individual rapidly show signs of recovery. For an individual with VWF, recovery takes longer. It is convenient to use the time for finger temperatures to recover by 4°C as a measure of vascular function. If the 4°C recovery

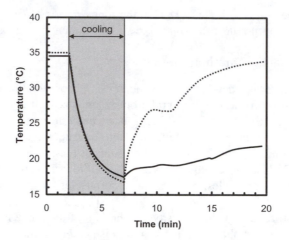

FIGURE 4.6 Finger temperature profiles for digits during cooling and recovery. Recovery is more rapid for a normal subject (······) than for a subject with VWF (——).

takes longer than 5 min, then damage is possible; if the 4°C recovery takes longer than 10 min, then damage is likely.

4.4.2.4   Finger Systolic Blood Pressure Testing

The finger systolic blood pressure (FSBP) is measured using a plethysmograph, parts of which operate on a similar principle to a sphygmomanometer (the device used to measure systemic blood pressure). FSBPs are measured after cooling the fingers to 15, 10, or 6°C. These results are compared to reference data obtained while fingers have been maintained at 30°C. The two FSBP measures are expressed as a percentage (FSBP%), after correction for the systemic blood pressure (Bovenzi, 2002; ISO/DIS 14835-2, 2004).

Each test of FSBP is carried out after cooling a finger that has been emptied of blood and occluded, using a pressure cuff. After 5 min of simultaneous cooling and occlusion of blood, the pressure is gradually released until blood flow returns to the finger, often measured using an annular strain gauge placed around the fingertip. The FSBP is the pressure at which the blood flow returns to the finger.

Testing can be carried out at 30 and 15°C. If an expected response is not obtained at 15°C, then the test can be repeated at 10°C. If the FSBP is less than 80%, then damage is indicated; if the FSBP is less than 60%, then damage is likely (Lindsell and Griffin, 1998).

4.5   REDUCING RISKS FROM HAND-TRANSMITTED VIBRATION

Effective health risk management requires a multifactorial approach involving many tiers of a company. Some guidance can be found in the literature (e.g., ISO 5349-2, 2001); alternatively an action flow diagram could be used (Figure 4.7). The most

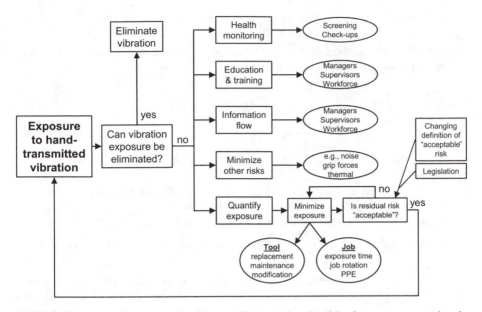

FIGURE 4.7 A possible action flow diagram for reducing the risks from exposure to hand-transmitted vibration.

effective way of reducing risk is to eliminate exposure to the vibration. Some industries with a vibration problem are in economic decline (e.g., U.K. coal mining) which has resulted in fewer individuals being exposed, but unemployment can hardly be recommended as a solution! Other industries continue to develop automated processes which require fewer "hands-on" tasks but still call for operators to supervise and control the machine. For some applications, remote-controlled tools can be used (e.g., vibratory compactors). If vibration exposure cannot be eliminated, then a range of other actions should be implemented in parallel. It is important to balance these actions across managerial, technical, and medical aspects.

A risk assessment of exposure to hand-transmitted vibration is not a one-time occurrence. Tasks change, workers change jobs, legislation changes, technologies improve, and methods for minimizing risk improve. The diagram in Figure 4.7 includes feedback to ensure that the assessment is repeated at appropriate intervals.

4.5.1 HEALTH MONITORING

Health monitoring should start at recruitment with preplacement screening and continue through regular check-ups. It is wise for employers to have some indication of previous vibration exposure and preplacement symptoms of Raynaud's disease, whether primary or secondary. Usually, a structured history questionnaire gives sufficient information, although some employers might decide to use preplacement physiological tests.

Symptoms of HAVS should be included as part of regular workforce health checks. Records should be kept so that any progression of symptoms can be noted. At a managerial level, a clear policy should be in place that defines what action to

TABLE 4.5
Possible Use of Method of Scoring Vibration-Induced White Finger to Assist in the Management of Health Risk

Maximum score on either hand	Possible action
0	No action required but continue to regularly monitor vibration-exposed worker for symptoms
1	Vibration-exposed person to be regularly warned of problem and informed of possible consequences
3	Vibration-exposed person to be advised not to continue with vibration work
5	Vibration-exposed person to be removed from vibration work
9	Compensation payable to vibration-exposed worker

Source: Adapted from Griffin, M.J. (1990). *Handbook of Human Vibration.* London: Academic Press.

take should a worker develop early stages of HAVS. In some cases, redeployment is the most appropriate; managers need to decide whether such action is advisory or mandatory. Griffin's blanching symptom score (see Subsection 4.4.1) can be used as an incremental system for decision making, depending on the vibration-exposed worker's symptoms on their most affected hand (Table 4.5).

4.5.2 EDUCATION AND TRAINING

All tiers of a company can benefit from training aimed at reducing the risk from hand-transmitted vibration. Managers have a responsibility to be aware of guidance documents, changes in legislation, and the implications of such changes. Supervisors should be trained in what to look for in terms of health effects or what might constitute a hazardous exposure. Tool users should be trained in how to operate and maintain the tool properly, minimizing health risks. In particular, when new low-vibration tools are introduced, workers should be educated to ensure that they expect the tools to feel different from those they are used to. (There are anecdotal reports across many industrial sectors where new tools have been unpopular due to opera-tives considering them less effective because of the reduced feelings of vibration at the handle.) The importance of maintenance should be emphasized, and for drilling and cutting tools, prompt replacement or sharpening of blunt cutting surfaces should be encouraged. An appreciation of the health risks at all levels should provide motivation for development of, and adherence to, company health and safety policies.

4.5.3 INFORMATION FLOW

It might be argued that continuous "information flow" is not strictly a policy to be taken for reducing health risks. However, in many practical situations, an impaired information flow can be a significant obstacle to effective risk reduction. Effective information flow is a characteristic feature of a genuine safety culture; the latter cannot be regulated but must be developed. Employees should feel free to approach their supervisors or health representatives to discuss their concerns, confident in the knowledge that support will be forthcoming whatever the outcome. Similarly, supervisors must be confident that worries on the shop floor are taken seriously at senior levels. Information flow should also be effective from the top down. When action (or a decision not to take action) is taken, the workers should be informed, especially if the action is in response to a specific query. In many cases, a healthy working relationship with trade unions can facilitate effective information flow.

4.5.4 MINIMIZATION OF OTHER ERGONOMIC RISK FACTORS

Many aspects of HAVS are likely to be due to tool use in general, rather than the vibration. Although, of course, vibration emission is a crucial element of the tool selection process, other aspects should not be neglected. It is not acceptable, for example, to have employees who are not at risk of VWF, but are very likely to develop industrial deafness. Other important factors include the design of the trigger and handle so that there is not a pressure point that might cause occlusion of blood flow in the finger; the weight of the tool should be minimized or supported on articulating arms or hoists. Any air outlets should be positioned such that the hands are not cooled; the tool must do the job effectively with minimal physical exertion of the operator (e.g., grip or push forces).

4.5.5 QUANTIFICATION AND MINIMIZATION OF EXPOSURE

In most situations where an operator is exposed to hand-transmitted vibration, the exposure should be quantified and compared to current guidance and legislation. In all cases, exposure should be minimized as far as is reasonably practicable. What constitutes "reasonably practicable" depends on the tool type and the magnitude of the vibration exposure. Inevitably, there will be some residual vibration exposure, and this must again be assessed to ensure that the residual risk is acceptable. Acceptability is a changing concept, and what might be considered acceptable in one country might be different from that in another. Similarly, with improvements in vibration reduction technology, better understanding of the damage mechanisms, and changing cultural trends, the recommended or legislative limits might be tightened.

There are two ways of reducing vibration exposure: either by optimizing the tool or by optimizing the job.

4.5.5.1 Optimization of the Tool

Some vibration problems are an issue of maintenance. For example, a rotary component that is out of balance generates vibration; a blunt cutting edge requires more time before the cut is complete. Well-maintained tools generally vibrate less and are more effective.

If excessive vibration exposure is not resolved by maintenance, then it might be possible to modify the tool. Some vibration-reducing handles have been developed, but purchasers should be cautious, confirming the durability and effectiveness of the handles for the tools in their application. Autobalancing devices can be fitted to some grinders, automatically compensating for damaged grinding disks (e.g., Åresk-oug et al., 2000).

Inevitably, tools must be replaced when they cease to be productive. Fortunately, good ergonomics is becoming more of a selling point for hand tools. With an increased volume of legislation (and litigation), purchasers are now more likely to actively seek out tools with low vibration emission. For many devices, there have been significant improvements in their modern versions. Suppliers must now provide declared values for tools to be sold within Europe (see Subsection 8.4.2), and freely accessible databases are available electronically (e.g., Mansfield et al., 1998). There-fore, purchasers should be able to use vibration emission as a tool-selection variable, although it must be stressed that a declared vibration emission might not be repre-sentative of an operator's vibration exposure (see Section 7.3).

4.5.5.2 Optimization of the Job

If the vibration exposure is such that it exceeds agreed acceptable limits for a specific duration of use, then using the tool for less time enables compliance to be achieved. This could be done by changing the task so that the tool does not need to be used for as long or by spreading the risk among workers by utilizing job rotation strategies.

For many tools, a "trigger time" (sometimes referred to as "anger time") of a few hours per day corresponds to a full working day due to other processes within the task (e.g., placement of workpiece, checking, alignment, delivery, collection, setting up, breaks). In these cases, even if the tool can only be used for a short time, the time allowable on the job could still be acceptable.

4.5.5.3 Antivibration Gloves

One might consider antivibration gloves as an ideal form of personal protective equipment (PPE). Unfortunately, there are many pitfalls with the design and selection of such gloves. Due to the nature of isolating materials, all gloves are more effective at isolating high-frequency vibration than low-frequency vibration, but it is the latter that is considered to be more hazardous, according to standardized assessment techniques that use the W_h weighting. Many early gloves amplified vibration at those frequencies at which the hand is most sensitive! In an effort to remove such sub-standard "antivibration" gloves from the market, ISO 10819 was introduced in 1996, providing a standardized test procedure with acceptance criteria (see also Subsection 7.4.2). For a glove to be classified as an antivibration glove, it must be tested in the laboratory, using two vibration stimuli. The M stimulus (medium frequencies) con-tains vibration in the frequency range of 16 to 400 Hz; the H stimulus (high frequencies) contains vibration in the frequency range of 100 to 1600 Hz. To pass the test, the ratio of the frequency-weighted accelerations measured inside the glove with a "palm adapter" to that on the test handle surface must be less than 1 for the M stimulus and less than 0.6 for the H stimulus. Therefore, at the low frequencies at which the first prototype antivibration gloves amplified vibration, the glove must on average attenuate, and the attenuation must be at least 40% for the high frequen-cies. Initially, very few gloves passed this standard; more recently, a range of gloves has been developed with improved dynamic performance so that the standard can be passed, and consequently these gloves can be sold as antivibration gloves.

One problem with ISO 10819 is that gloves can be designed to pass a test, rather than to protect a worker. For example, some gloves amplify vibration over a range of frequencies in the M spectrum but, because they attenuate at other frequencies, the mean transmissibility remains less than 1. Hence, if the glove is used with a tool whose vibration is dominated at the glove's resonance frequency, then the glove acts as an amplifier rather than a piece of protective equipment. Some more complex methods have been proposed that can overcome this problem of testing by using techniques similar to the SEAT value (see Subsection 2.5.2) for seating (e.g., Griffin, 1998; Rakheja et al., 2002). All of these test methods measure at the palm of the hand rather than at the fingers where the symptoms of VWF occur, but where it is also far more difficult to reliably measure the vibration. Other gloves that comply with ISO 10819 require thick layers of isolation material which reduce the efficiency of the grip and flexibility of the hands, leading to poor gripping postures. Finally, there can be a problem with durability for some gloves in extreme vibration environments where the vibration can destroy the isolation material after just a few days use.

ISO 10819 states that:

> A glove shall only be considered as an antivibration glove according to this standard if the fingers of the glove have the same properties (materials and thickness) as the part of the glove covering the palm of the hand.

Some gloves are available that are fingerless and, according to this clause in the standard, should not be termed antivibration gloves. Fingerless gloves cannot be expected to be effective at reducing risks as the bare fingers remain in contact with the vibrating tool.

The question of whether or not to recommend the use of antivibration gloves is difficult to answer. Each case must be taken on its merit, ideally matching the dynamic properties of the glove to the vibration characteristics of the tool. Careful attention should be given to the possible introduction of other ergonomic risk factors, and the long term performance and durability of the glove should be closely monitored. If glove technology continues to improve, then antivibration gloves could provide clear benefits to those at risk of HAVS.

4.6 CHAPTER SUMMARY

Hand-transmitted vibration is perceived across a wide frequency range due to the combined responses of a range of receptors located in the skin. Experimental investigations of the sensitivity of the fingertip to vibration have enabled the W_h frequency weighting to be developed, which is used to model the response of the hand to vibration. If an individual is exposed to hand-transmitted vibration for extended periods of time, then he or she is at risk of developing HAVS that is characterized by vascular and nonvascular damage. The best known component of HAVS is VWF, which is a form of secondary Raynaud's disease. This presents as an episodic blanching of the fingers, which first affects the fingertip but spreads if vibration exposure is not curtailed. Attacks of VWF are triggered by exposure to cold and might not be observed in the workplace where the damage is usually done.

Considering the high stakes in an increasingly litigious society, it is important that HAVS is diagnosed and classified reliably, which is difficult considering its episodic nature. A variety of diagnostic methods can be used ranging from classification according to the Stockholm scale or using Griffin's method of scoring the extent of injury to using a battery of objective diagnostic techniques involving specialized clinical equipment.

The risks of vibration injury should be minimized using managerial, medical, and technical measures. If it is not possible to eliminate exposure completely, then action should be taken to reduce the risks, and there should be regular reviews. Exposure can be minimized by improving the tool or by improving the job design (e.g., job rotation). Although antivibration gloves are available, their effectiveness is far from being established, but they might form part of a hand-transmitted vibration-risk-reduction strategy.

5 Vibration Measurement

5.1 INTRODUCTION

When we approach a human vibration problem, it is usually helpful, and sometimes mandated, to make an objective measurement. Vibration measurement is a complex topic and there are many paths that can be taken to turn a mechanical motion into a value or figure in a report. Some of these paths might be valid; others might not be valid; still others might be useful indicators, but their validity has not been established in the research arena. Even some methods apparently advocated in certain standards have not been validated! One of the problems with vibration measurement is that even if an incorrect method has been used, most measuring equipment can still generate a number on a display. A nonexpert has no way of knowing whether the measurement has been a success or not.

Unfortunately, it is impossible to gain experience by reading any chapter of any book. Experience can only be gained by getting on site, using the equipment, and overcoming the practical challenges that are inevitable when measuring in the challenging environments that usually accompany a vibration problem. By definition, the unpredictable cannot be predicted, and it is the experience of the author that unpredictable events are common when it comes to making vibration measurements.

A full understanding of the function of each element in a vibration measurement system and the system integration will empower an investigator to avoid erroneous data. One attraction or repulsion (depending on the individual's point of view) to becoming informed regarding measurement is the broad cross section of physical principles, electronics, computer technologies, digital signal processing, analog signal processing, mathematics, and mechanical engineering that this discipline involves. Fortunately, for a nonengineer, all of the component parts required to make sophisticated vibration assessments are available off the shelf and simply require connection and some training in their use. However, it is still important that at least the basic principles behind the technology are grasped so that repeatable and reliable measurements are made while avoiding the pitfalls.

This chapter provides practical guidance on the measurement of human vibration. An overview of the measurement process is provided in Section 5.2. The two options for the measurement strategy are discussed in Section 5.3 and Section 5.4. The function and operation of each component part of a measurement system are covered in Section 5.5 to Section 5.8. Steps that can assist in making successful measurements are introduced in Section 5.9, and the chapter concludes with two case studies: a hand-transmitted vibration assessment (Section 5.10) and a whole-body vibration assessment (Section 5.11).

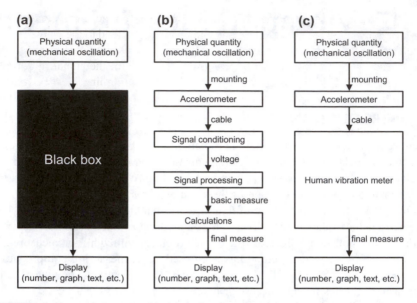

FIGURE 5.1 Component parts of human vibration measurement systems represented as (a) an idealistic "black box" solution, (b) a modular system constructed from component parts, and (c) a system using an integrated human vibration meter.

5.2 MEASUREMENT FLOW

Between the mechanical oscillation and the final numeric values forming a measurement, there are a series of steps. The simplest way of considering these steps is to group them into a "black box" with an input of the physical quantity and the output of the desired measurement presented on a display (Figure 5.1a). If we dismantle this utopian black box, then we will find that it must consist of a transducer, some signal conditioning, and at least some basic signal processing.

A practical system is illustrated in Figure 5.1b. First, a transducer must be attached to the vibrating surface. The transducer (almost always an accelerometer) converts the mechanical oscillation into an electrical property. For example, acceleration might cause the electrical resistance of the accelerometer to change. The signal conditioning, which is attached to the transducer by a cable, converts the electrical property into a voltage that can be processed. Modern vibration-measuring equipment converts the voltage into a digital signal on which complex calculations can be carried out.

When a human vibration assessment is made, the final "measurement" depends on a reliable flow of the signal. This includes not only the main components of the system but also the connections. Mistakes can be made even before the vibration affects the electromechanical properties of the accelerometer if it is not mounted to the vibrating surface adequately, of if the wrong surface is measured. The accelerometer must be appropriate so that the important frequencies are measured (low-frequency vibration should be measured for motion-sickness assessments; intermediate frequencies should be measured for whole-body vibration assessments; high frequencies should be measured for hand-transmitted vibration assessments). A good quality cable should be used to connect the accelerometer to the signal conditioning; otherwise, unwanted electrical noise can be inadvertently added to the wanted signal, reducing the fidelity of the measurement. Also, care should be taken to reduce any unnecessary cable movement as this can also interfere with the signal due to phenomenon known as the "triboelectric effect." Signal conditioning must be matched to the accelerometer type, and signal processing must be matched to the signal conditioning. For example, if the signal conditioning produces a voltage of 10 V, then the signal processing must be capable of measuring 10 V. Most of the signal processing and calculations to apply the results to standardized methods is usually completed within the software. Finally, the display should be well designed such that it is usable for the operator of the equipment.

Although it is possible to combine some elements of vibration-measuring components into "single-box" solutions (e.g., a human vibration meter; Figure 5.1c), these still contain the same components internally as the full expanded system. The user must still ensure that each part is set up appropriately, which can mean navigating through multilayer menus within the device software.

5.3   HUMAN VIBRATION METERS

An increasing number of "human vibration meters" have come onto the market since the 1990s. Their development has been prompted by an increased demand for basic vibration measurement, largely in response to the development and implementation of the European Union Physical Agents (Vibration) Directive. Some meters are designed for assessment of hand-transmitted vibration only, others are more generally applicable. Many are based on sound-level-meter technology and it is possible to purchase a single device that can be simply converted from a sound-level-meter to a hand-transmitted vibration meter. The advantage of a human vibration meter over a modular system is that it is a relatively simple device giving an immediate measure of the vibration magnitudes without the requirement for analysis in the laboratory. Naturally, the lack of complex analysis techniques in the meter itself restricts its use to those situations where a simple magnitude measure is adequate. Most meters are not able to store the acceleration waveform, and so the user is unable to examine the nature of the signal and the fidelity of the measurement. It is usually necessary to not only identify that a problem exists but to understand why the problem exists, so that solutions can be sought.

Although meters can be easy to use, the user must still ensure that the device is correctly configured. For example, frequency weightings, axis multipliers, calibration factors, analysis methods, and a host of other possible variables must all be programmed appropriately for a valid assessment to be carried out. For basic measurements, meters might be simpler than modular systems, but there is still scope for methodological errors.

As the waveform is not stored on the meter, it is impossible to investigate why some measures might be abnormally high or low. Care must be taken to avoid measuring nonvibration artifacts such as mounting or repositioning of the accelerometers or events such as drivers occupying or leaving the seat (e.g., Atkinson et al., 2002).

5.3.1 Considerations for Selection of a Human Vibration Meter

When sourcing a human vibration meter, a number of factors should be considered. All meters should be shown to comply with ISO 8041 (1990), which specifies minimum requirements for the instrument. The meter should include the appropriate frequency weightings: at least W_h for hand-transmitted vibration; W_d, W_k, and possibly W_b for whole-body vibration; and W_f for motion sickness. If the meter is only required for one type of assessment, then it is unnecessary to purchase a device with frequency weightings suitable for all types of environments.

Some meters can measure more than one direction of vibration simultaneously; others are restricted to making multi-axis assessments by sequentially measuring each direction. Standards require triaxial assessments to be made for both hand-transmitted and whole-body vibration. Therefore, a multichannel meter capable of measuring at least three axes simultaneously can substantially reduce the time required to test. Additionally, for nonstationary signals (i.e., those whose statistical properties change over time), it is not appropriate to measure each direction separately. Often, the capability to measure six channels simultaneously is practically helpful: for hand-transmitted vibration assessments, this allows for measurement of both hands at the same time; for whole-body vibration assessments, measurements can be made on the surface of a seat as well as at the base of a seat, allowing for calculation of the SEAT value.

Functions allowing computer interfacing might be attractive to some users. For example, it might be possible to interface a meter to a computer for transfer of data, programming, updating internal software, and backup. These features might often go unused by some investigators; others might regard them as essential. If the operator makes use of internal memory, then the memory size might be important. Some such devices allow for connection of the meter to data loggers so that the raw signal-time history can be acquired onto a computer. In this configuration, the meter simply acts as a signal-conditioning unit.

If the purchaser of a human vibration meter already owns some accelerometers, then it might be possible to source a compatible instrument. Alternatively, accelerometers might be required to be supplied as part of the package. In this case, the suitability of the accelerometers should be checked. For example, if measurements are to be made on a seat surface, then the accelerometer should fit within the standard seat-testing pad or be supplied already mounted within a pad.

Finally, the meter must be usable by the operator. Ergonomic principles for good software design can be overlooked by some manufacturers. When considering an instrument for purchase, hands-on experience can be valuable for making an assessment of the ease of use. All common tasks should be simulated (including setting up, calibration, measurement, and data transfer).

5.4 DATA-ACQUISITION SYSTEMS

If a system more flexible than a vibration meter is required, then a modular data-acquisition system can be assembled. Within a modular system, the output from the accelerometer is converted to a voltage signal using appropriate signal conditioning. This voltage is the property that the system can measure and analyze.

Data-acquisition systems are usually computer based, although some stand-alone data loggers are available. These have the capability of sampling a voltage (i.e., the conditioned acceleration) at discrete time intervals such that the waveform of the voltage can be stored on a computer. The stored waveform can then be analyzed using any compatible software. If any anomalies have occurred within the measurement, then these can be viewed to help provide an explanation. Usually, devices are capable of sampling on more than one channel simultaneously; many are 8- or 16-channel systems, allowing for 8 or 16 vibration measurements to be made simultaneously. For each triaxial measurement, three channels are required.

Most computers require an additional interface to measure voltage with the precision required for vibration analysis. These interfaces can be internal (e.g., PCI cards and PCMCIA cards) or external (e.g., USB, parallel, and serial). With improvements in computing technologies, many systems previously limited to laboratory use are now usable in the field. This trend of improved power is likely to continue with higher specification communications protocols being exploited. However, essentially, the same phenomena are being measured; acceleration is converted into a voltage and the voltage is converted to digital data, whether the computer is a large desktop machine or a small handheld. Historically, one problem with data-acquisition systems is that large files are easily created and available memory could be a problem (some of the first systems shared the program and all data files on one floppy disk!). As a result, measurement durations could be limited. Another practical issue is that most computers used for data acquisition are not designed to be used in harsh environments and so reliability can be a problem if the computer must travel with the tool or machine rather than be connected via a wireless link or cable.

5.4.1 RESOLUTION AND SAMPLING

The two key parameters to consider when configuring a data-acquisition system are the resolution and the sample rate. The resolution is dictated by the number of bits per sample. The most common systems are 8-bit, 12-bit, 16-bit, or 24-bit. As the number of bits increases, so the precision of the sampled signal improves, but the memory required to store the data increases. It is generally considered that 12-bits are the minimum required for human vibration data-acquisition systems.

The signal is repeatedly "sampled" and each sample is converted to a binary number by an analog-to-digital converter (ADC). The length of the binary number is prescribed by the number of bits. For example, a single-bit number can be either 0 or 1; a 2-bit number can be 00, 01, 10, or 11. Each additional bit doubles the number of possible values or "states" that the ADC can assign. For the single-bit system, there are 2^1 possible states (i.e., 2); for the 2-bit system, there are 2^2 possible states (i.e., 4). Similarly, a 12-bit system can measure a possible 2^{12} (i.e., 4096) states, a 16-bit system can measure a possible 2^{16} (i.e., 65,536) states, and a 24-bit system can measure a possible 2^{24} states (i.e., over 16 million possible values). In principle, the ADC selects the state that is closest to the value of the measured voltage. Figure 5.2 illustrates how the sampled signal might be measured for 1-, 2-,

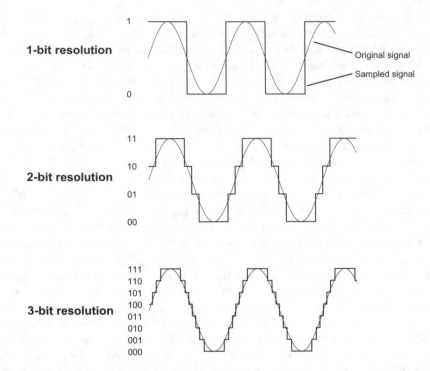

FIGURE 5.2 Illustration of the improvement in digitized waveform fidelity as the number of bits per sample increases from one to three. Human vibration data-acquisition systems use at least 12 bits per sample.

and 3-bit converters. The sampled signal for the 1-bit ADC bears little resemblance to the original signal; as the characteristics of the ADC improves, the fidelity of the sampled signal improves.

The ADC samples the voltage at its input at discrete time intervals. The faster the sample rate, the shorter the interval between each measurement. If the system samples at one sample per second, then the voltage is repeatedly measured each second. For most human vibration analyses, measurements are made many hundreds, or thousands, of times per second to ensure that a complete picture of the vibration environment can be determined. Figure 5.3 illustrates the importance of selecting the correct sample rate. If the rate is too slow, then the signal that is to be measured cannot be reproduced from the sampled data. There is a threshold where the general characteristics of the signal is retained. Ideally, none of the information in the original signal is lost, with many samples being quantified per cycle. As a rule of thumb, the minimum sample rate should be at least three times the highest frequency of interest in the signal. If the equipment is capable, then a rate of 10 times the highest frequency of interest should be selected. For example, if the highest frequency of interest is 100 Hz, then each accelerometer should be sampled at least 300 times per second, and ideally at least 1,000 times per second.

Usually, sample rates are selected to be a power of two (e.g., 256, 512, 1,024, or 2,048 samples per second). This enables convenient resolutions to be selected

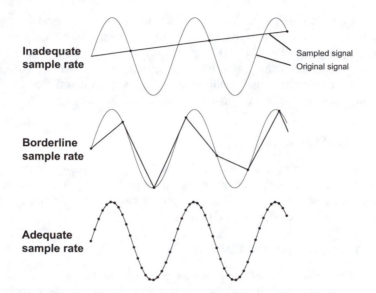

FIGURE 5.3 Effect of sample rate on the fidelity of a digitized signal acquired by a data-acquisition system. As the sample rate increases, the sampled signal bears more resemblance to the original signal. If the rate is too low, then the original signal will not be measured by the data-acquisition system.

when analyzing in the frequency domain, such as reporting data every 0.25, 0.5, or 1 Hz. If other sample rates are selected, then inconvenient frequency resolutions are imposed. For example, if data are acquired at 100 samples per second, then analysis in the frequency domain must use resolutions such as 0.195, 0.391, and 0.781 Hz. It would thus be possible for a reader of a report to misinterpret measurements as having a precision greater than is really the case (or is desirable in most human vibration contexts). This error is less likely to occur if measurements are reported with a simpler frequency interval.

5.4.2 ALIASING

If there is a signal present that cannot be sampled adequately, it can compromise all data being measured. This has the effect of appearing as a lower frequency signal in the sampled data, an effect known as aliasing (Figure 5.4). Unless precautions

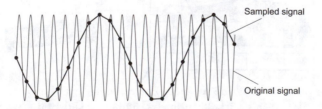

FIGURE 5.4 A sampled signal showing aliasing. The sample rate is inadequate to measure the original signal. As a result, the sampled data indicates a low-frequency "aliased" wave that does not exist in the original signal.

are taken, it is impossible to know whether any measured signal is a genuine acceleration or an aliased high-frequency signal. Any signal with a frequency above half the sampling rate of the ADC (known as the Nyquist frequency) will be aliased. The solution to the problem of aliasing is to use high sample rates for the ADC and antialiasing filters in the signal conditioning. The low-pass cutoff frequency of the antialiasing filter should be less than one third of the sample rate.

5.5 ACCELEROMETERS

An accelerometer is a device in which electrical properties change in proportion to the acceleration to which they are exposed. There are three types of accelerometers: piezoresistive, piezoelectric, and ICP (integrated circuit piezoelectric). Each type requires its own type of signal conditioning.

5.5.1 PIEZORESISTIVE ACCELEROMETERS

Piezoresistive accelerometers use strain gauges to sense the acceleration, and are therefore sometimes referred to as strain gauge accelerometers. These strain gauges are configured as a Wheatstone Bridge electrical circuit bonded to a beam that is fixed to the accelerometer casing at one end and a seismic mass at the other end (Figure 5.5). When the transducer is exposed to an acceleration, the inertia of the seismic mass generates a bending force in the beam. The resulting bending causes strain in the strain gauges, which can be converted to a voltage using a strain-gauge amplifier. This strain (and therefore output voltage) is proportional to the acceleration.

Gravity always acts on the seismic mass and therefore the beam bends depending on the inclination of the piezoresistive accelerometer. As a result, the output for a vertically inclined accelerometer provides a measure of +1g acceleration; the output for an inverted accelerometer provides a measure of –1g acceleration. As gravity is nominally constant across the earth's surface, this property can be used for calibration purposes and for checking the orientation of the accelerometer. Due to the principle of operation, piezoresistive accelerometers are suitable for the measurement of low-frequency vibration. Their performance is usually limited at high frequencies. There

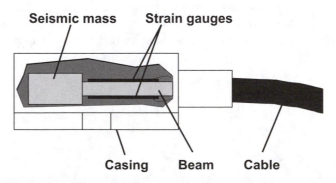

FIGURE 5.5 Cross section of a piezoresistive accelerometer.

is a trade-off between accelerometer sensitivity, robustness, and the high-frequency response. The more the beam inside the transducer bends, the more sensitive the accelerometer is. However, the more it bends, the less robust it is, and the internal resonances of the accelerometer occur at a lower frequency.

Considering the frequencies over which piezoresistive accelerometers work optimally, they are usually only used for motion sickness and whole-body vibration and are rarely suitable for hand-transmitted vibration applications.

5.5.2 PIEZOELECTRIC ACCELEROMETERS

Piezoelectric accelerometers use crystals whose electrical charge changes with the acceleration, and are therefore sometimes referred to as charge-type accelerometers. The piezoelectric crystal is fixed to the base of the accelerometer on one side and to a seismic mass on the other side (Figure 5.6). When the transducer is exposed to an acceleration, the inertia of the seismic mass generates either a compression or extension force or a shear force on the crystal. This force causes the crystal to produce a charge output (measured in coulombs) proportional to the acceleration due to the piezoelectric effect. This charge is converted to a voltage using a charge amplifier. Most modern piezoelectric accelerometers are configured to produce shear forces on the crystal.

Piezoelectric accelerometers are unable to measure continuous or very low-frequency acceleration. Therefore, they are not sensitive to gravity in the same way as are piezoresistive accelerometers and must be calibrated using a known vibration source. Piezoelectric accelerometers are limited at low frequencies. Therefore, the user must take care to ensure that the accelerometer can measure at the lowest frequency of interest. These accelerometers perform well at high frequencies and some can be extremely robust. Many are suitable for measuring impact accelerations.

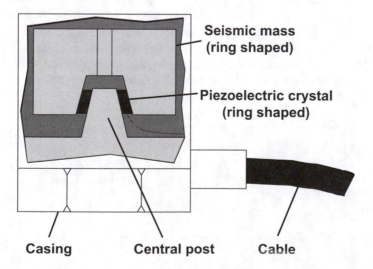

FIGURE 5.6 Cross section of a piezoelectric accelerometer.

Some piezoelectric accelerometers can be used for whole-body vibration applications, but not all. Others are suitable for hand-transmitted vibration applications.

5.5.3 ICP (INTEGRATED CIRCUIT PIEZOELECTRIC) ACCELEROMETERS

ICP are the most common of a range of "chip-type" accelerometers. They can look like conventional silicon chips that can be mounted on a circuit board. Others come premounted and cased such that they appear similar to the traditional piezoelectric and piezoresistive accelerometers. Inside the chip itself is a piezoelectric element with appropriate electronics such that the output is already converted to a voltage. Therefore, some accelerometers simply require a power source and can give a direct voltage output proportional to the acceleration. Others are configurable according to the properties of the electronic components that the user can connect to the other pins of the chip. The technical advantage of ICP accelerometers is that the required signal conditioning is simpler. A practical advantage is that they are usually less expensive than equivalent piezoelectric transducers. However, the user should be cautious because ICP accelerometers are manufactured for a broad range of applications. As a result, many low-cost devices are not appropriate for human vibration measurements due to imprecision, nonlinearities, and instabilities.

5.6 SIGNAL CONDITIONING

5.6.1 ACCELEROMETER AMPLIFIERS

Signal conditioning converts the output from an accelerometer into a voltage that can be measured by a data acquisition and analysis system. Piezoresistive accelerometers require a strain-gauge "bridge" amplifier and piezoelectric accelerometers require a charge amplifier. ICP accelerometers often only require a power supply. The output of the signal conditioning must be compatible with the data-acquisition system. For example, if the system is capable of measuring from −5 to +5 V, then the output from the amplifier should not exceed ±5 V; otherwise, clipping will occur (Figure 5.7). If the output from the amplifier is too small, then internal electrical noise in the system might become significant and the data will not reflect the full capabilities of the system to make precise measurements.

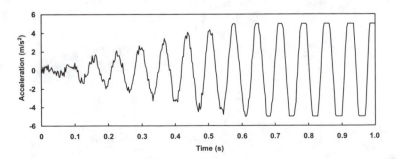

FIGURE 5.7 The effect of clipping on a noisy sinusoidal acceleration signal with increasing vibration magnitude. The signal is clipped when it reaches ±5 m/s^2.

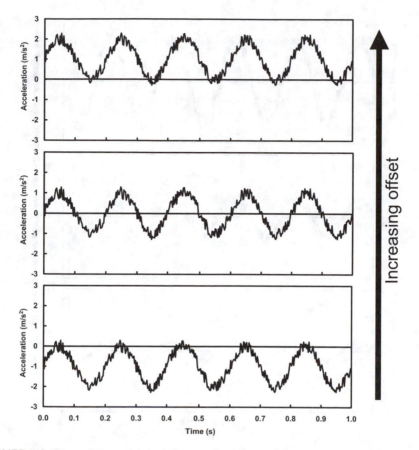

FIGURE 5.8 Output from a piezoresistive accelerometer exposed to a noisy sinusoidal acceleration. Effect of changing the offset of the strain-gauge amplifier.

Strain-gauge amplifiers have two adjustments: an "offset" and a "gain." When a piezoresistive accelerometer is plugged into the amplifier and the accelerometer is aligned to provide zero output, the offset adjuster can be used to set the output to zero (Figure 5.8). The gain control adjusts the sensitivity of the amplifier such that the output can be matched to the input of the ADC and the sensitivity of the accelerometer (Figure 5.9). Adjusters found on strain-gauge accelerometer amplifiers do not usually have numeric indicators, and all adjustments are carried out using a numerical display to set the correct adjustments.

Charge amplifiers must be set according to the sensitivity of the accelerometer as provided by the manufacturer. The output of the amplifier must be set according to the characteristics of the ADC. For example, an accelerometer might have a sensitivity of 1.01 pC per m/s^2 and needs to be connected to an ADC with a range of ± 1 V. In this case, the charge amplifier must be set to reflect the sensitivity (1.01 pC per m/s^2) and the output amplifier set such that the maximum voltage will not exceed ± 1 V. If the maximum acceleration is expected to be not more than, for example, 6 m/s^2 then the output amplifier could be set to 100 mV per m/s^2 such that

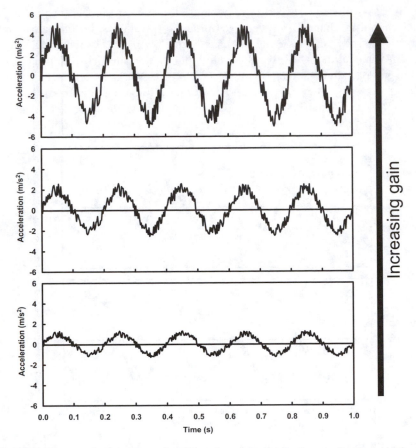

FIGURE 5.9 Output from a piezoresistive accelerometer exposed to a noisy sinusoidal acceleration. Effect of changing the gain of the strain-gauge amplifier.

the maximum output voltage will not exceed 600 mV. In this case, the maximum acceleration could be slightly higher than expected without exceeding the limits of the ADC.

5.6.2 FILTERS

One of the most important elements of signal conditioning is the presence of high-pass and low-pass filters. A filter removes elements of the signal that are not of interest. A high-pass filter cuts low-frequency elements below the user-selectable cutoff frequency; a low-pass filter cuts high-frequency elements above the user-selectable cutoff frequency. Filters are sometimes combined and referred to as band-pass filters. The characteristics of the filter can be described by the number of poles or the decibels per octave (dB/octave). Essentially, the better the filter is at removing unwanted data, the greater the number of poles and the greater the number of dB/octave.

High-pass filters are useful in removing artifacts from signals such as accelerations caused by changes in speed and cornering in vehicles. For human vibration analysis, the most important filter is usually the low-pass or antialiasing filter. As aliasing occurs at the ADC (see Subsection 5.4.2), the filter must be placed in the system prior to digitization. Although signal-processing software includes digital high- and low-pass filters, these are not suitable for use as antialiasing filters, as the digitization has already taken place before they can be implemented. Aliasing is likely for all types of human vibration measurement and so precautions must be taken to ensure good quality measurements. It is good practice to set the antialiasing filter at less than one third of the sampling rate. Many, but not all, accelerometer signal conditioning units contain antialiasing filters.

5.7 CALIBRATION OF HUMAN VIBRATION MEASUREMENT SYSTEMS

The previously described process will allow for acceleration signals to be reliably converted to a voltage and acquired to a data-logging system. The user therefore has a measure of the vibration in terms of voltage. These voltages must be converted to acceleration to make meaningful interpretations of the data. This process must be accurate to provide a calibrated measurement.

There are two approaches to calibration: the first is to use the manufacturers' quoted sensitivity for acceleration; the second is to expose the accelerometer to a known acceleration source and to scale the signal accordingly. For piezoelectric accelerometers, it is often appropriate to combine these approaches, using the manufacturers' data to set up the charge amplifier and to check the setup by using a dedicated calibrator. Piezoresistive accelerometers are calibrated using a known acceleration source.

Calibrators produce a known acceleration at a mounting point where accelerometers are fixed. Field calibrators are robust instruments that can be used on-site just prior to mounting of accelerometers. Many of these produce vibration at 159.2 Hz (1000 rad/s) at 10 m/s^2 root-mean-square (r.m.s.). A calibrated system should therefore indicate 10 m/s^2 r.m.s. when the accelerometer is attached to such a calibrator.

For accelerometers that are capable of measuring continuous acceleration (i.e., piezoresistive devices) the acceleration due to gravity is a convenient (and cheap!) known acceleration source. When the accelerometer is aligned so that it is sensitive in the vertical direction, it outputs +1g (9.81 m/s^2); when it is inverted, it outputs −1g (−9.81 m/s^2). It is usually convenient to use the offset control on the strain-gauge amplifier to set the vertically aligned output to 0 to maximize the effective dynamic range of the ADC. Therefore, when the accelerometer is inverted, it measures −2g (−19.62 m/s^2). Any horizontal surface can therefore be used as an *ad hoc* calibrator. Figure 5.10 shows a typical output from a piezoresistive accelerometer undergoing an inversion test. Initially, the output is zero when the accelerometer is placed on a horizontal surface and aligned such that it is sensitive in the vertical direction. When it is removed from the surface, the accelerometer responds to the movement and indicates a short peak. The signal decreases as the accelerometer is

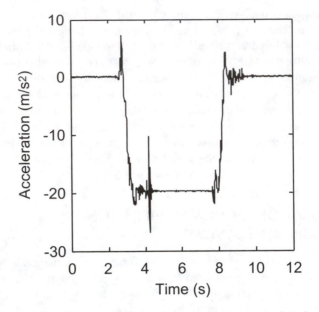

FIGURE 5.10 Typical signal from a piezoresistive accelerometer undergoing an inversion test. The accelerometer signal conditioning offset and gain are set such that when upright it measures 0 m/s^2, and when inverted, measures -19.62 m/s^2. Impulses in the signal correspond to impacts with the horizontal surface.

twisted and inverted so that it is upside down. When it is placed back on the horizontal surface, a small impact is measured followed by a continuous measure of -19.62 m/s^2. Finally, the accelerometer is returned to its original upright position on the horizontal surface and the output reads 0 m/s^2.

If the output from the accelerometer is not as expected or required, then the settings need adjustment either in the hardware (amplifiers) or in software (e.g., a multiplying and scaling factor). Alternatively, some part of the system might be faulty. One pitfall to avoid when calibrating accelerometers is to ensure that frequency-weighting filters are switched off during the calibration (and then switched back on again prior to the measurement). For example, the low-frequency accelerations from gravity and the relatively high-frequency accelerations from the example field calibrator will be filtered out by the band-pass filters that are standardized for whole-body vibration measurements.

5.8 SIGNAL PROCESSING

Once reliable and calibrated signals from the accelerometers have been acquired, they must be processed to generate numerical indicators suitable for the purpose of the assessment. A frequency weighting will almost always be required, as will basic statistical descriptors. Depending on the application, frequency analysis might be carried out.

5.8.1 FREQUENCY WEIGHTINGS

A frequency weighting provides a model of the response of a person to the vibration. People are more sensitive to some frequencies of vibration than others, and this frequency dependence is simulated using the frequency weightings. For example, the body is more sensitive to whole-body vibration at about 5 Hz than at 50 Hz; therefore, the vibration at 50 Hz is weighted (attenuated) such that its relative contribution to the total signal is reduced accordingly (see also Subsection 2.2.4 and Section 4.2). Frequency weightings are designed to not affect those frequencies where the body is most sensitive and to attenuate at those frequencies where the response of the body is less sensitive. In principle, weightings do not amplify at any frequency. Therefore, the magnitude of the frequency-weighted signal should not be more than the magnitude of the unweighted signal.

Frequency weightings can be implemented using analog electronics or, most commonly, using digital signal processing techniques. To be able to calculate all statistical functions, the weighting should be carried out in the time domain (i.e., it should be applied to the time history of the acceleration signal). Implementation of frequency weightings is difficult, and usually specialist software is required. However, time-domain weightings can be written using a variety of techniques including multistage recursive filters or using convolution.

Although a wide variety of weightings can be found in the literature and in standards (Table 5.1), the most commonly used are W_d and W_k for whole-body vibration, and W_h for hand-transmitted vibration. A potential user selecting measuring instrumentation should ensure that at least these three weighting filters are provided.

TABLE 5.1
Summary of the Most Common Frequency Weightings Used for Analysis of Human Vibration Signals

Frequency Weighting	Application Area	Frequency Range	Direction
W_b	Whole-body	0.5–80 Hz	z-seat
W_c	Whole-body	0.5–80 Hz	x-backrest
W_d	Whole-body	0.5–80 Hz	x-seat
			y-seat
W_e	Whole-body	0.5–80 Hz	Rotation-seat
W_f	Motion sickness	0.1–0.5 Hz	z-vertical
W_h	Hand-transmitted	8–1,000 Hz	x-hand
			y-hand
			z-hand
W_k	Whole-body	0.5–80 Hz	z-seat

5.8.2 STATISTICAL MEASUREMENTS

For simple environmental assessments, the final stage of the measurement process is to carry out statistical measures of the frequency-weighted acceleration signal. Mathematically, these quantities are straightforward to calculate, and so all human vibration hardware and software solutions should include at least r.m.s., peak acceleration, crest factor and, if whole-body vibration analyses are required, vibration dose value (VDV) for each accelerometer.

5.8.2.1 Root-Mean-Square (r.m.s.)

By definition, vibration is a movement that oscillates about a fixed point. Therefore, assuming that there is no translation, the mean value of a vibration signal will, in theory, always be zero as all of the positive values will cancel out all of the negative values (if it is measured for an infinite duration or for a complete number of cycles at all frequencies). As such, the mean of the acceleration will not indicate the magnitude of the signal. The r.m.s. solves this problem by squaring every value in the signal, taking the mean, and taking the square root of this final value. r.m.s. is identical to standard deviation, if there is no offset in the acceleration signal. The unit of r.m.s. acceleration is m/s^2.

Some of the features of r.m.s. are illustrated in Figure 5.11 and Figure 5.12. In Figure 5.11 the acceleration signal is characterized as random vibration. In Figure

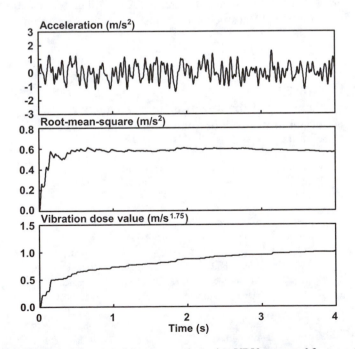

FIGURE 5.11 Acceleration, running r.m.s., and running VDV measured for a random signal. The r.m.s. rapidly iterates to a stable value, whereas the VDV continually accrues.

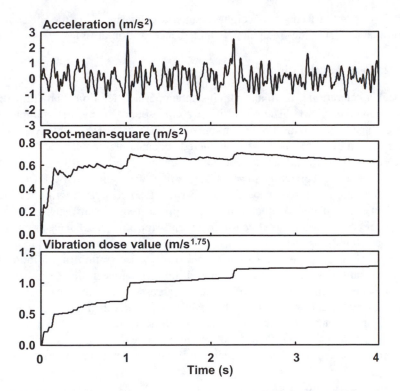

FIGURE 5.12 Acceleration, running r.m.s., and running VDV measured for a random signal with added shocks. The r.m.s. shows a decay in value following each shock, whereas the VDV continually accrues with steps at each shock.

5.12, the acceleration signal is characterized as random vibration (identical to that in Figure 5.11) but also containing two distinct shocks, one at about 1 s and one at about 2.2 s. During the first half second, the r.m.s. varies, but as the averaging period increases, the value stabilizes at about 0.6 m/s² r.m.s. For the random signal, the r.m.s. remains stable for the rest of the measuring period. For the signal containing shocks, the r.m.s. rapidly increases during each of these events, but this increase decays as the averaging time increases, with the r.m.s. tending towards the final value for the signal with no shocks. The final values for random vibration and for vibration containing shocks are 0.56 and 0.63 m/s² r.m.s., respectively. This example illustrates the main criticism against using r.m.s. for human vibration exposure assessments; it is relatively insensitive to occasional shocks because their influence decays as the measurement time increases. Therefore, if, for example, a driver is exposed to severe but occasional end-stop impacts, these might not be reflected in the r.m.s. despite these individual events being a major contributor to health risk.

Mathematically, r.m.s. can be expressed as:

$$a_{w\,r.m.s.} = \sqrt{\frac{1}{T}\int_0^T a_w^2(t)\mathrm{d}t}$$

where $a_{w\,r.m.s.}$ is the frequency-weighted r.m.s. acceleration, T is the measurement duration, and $a_w(t)$ is the frequency-weighted acceleration at time t.

5.8.2.2 Peak Acceleration

The peak acceleration is the maximum instantaneous acceleration at any time during the measurement period. Therefore, it only corresponds to the one sample of an acquired time history with the maximum value. For vertical seat vibration, the sign of the peak can indicate whether an untethered occupant momentarily left the surface of the seat at any time. A peak acceleration of less than $-1g$ (-9.81 m/s^2) would result in the seat accelerating towards the earth faster than gravity. Positive peak accelerations apply a force away from the earth and so occupants will not leave the surface of the seat unless other forces are applied (e.g., from the legs or arms). Occasional peaks in the signal can indicate end-stop impacts if measurements are being made on a suspension seat.

Some vibration meters report maximum r.m.s. acceleration in addition to the peak acceleration. There is a fundamental difference between these quantities: the maximum r.m.s. is the highest "level" averaged across a preset time averaging period (e.g., 1 s, usually corresponding to the update rate of the display); the peak acceleration is the highest instantaneous acceleration measured at any time during the entire measurement period. Peak acceleration will always be greater than the maximum r.m.s. acceleration. It is essential that these measures are not confused.

5.8.2.3 Crest Factor (CF)

The crest factor is a dimensionless quantity defined as the ratio of the peak acceleration to the r.m.s. The lowest possible crest factor is 1, which occurs for a square wave; a sine wave has a crest factor of 1.4; Gaussian random vibration has a crest factor of 1.7. If any of these signals contained a single instantaneous shock, then the crest factor would increase, but the r.m.s. might not be substantially affected. Therefore CF is useful in assessing the applicability of r.m.s. averaging. For the example vibration illustrated in Figure 5.11 and Figure 5.12, CF for the random signal is 2.8 and CF for the signal with shocks is 4.3. Using CF alone, without reference to the time history, these statistics indicate that the second signal contains more extreme shocks than the first, despite their similar r.m.s. magnitudes. For hand-transmitted vibration, CF might be of interest but of minor practical use, as r.m.s. is used for all standardized analysis methods. For whole-body vibration, CF is sometimes used to indicate whether r.m.s. or alternative assessment techniques are appropriate. The CF threshold is not universally agreed upon, but BS 6841 (1987) uses a value of 6 to designate where r.m.s. methods become invalid and where VDV should be used.

Mathematically, CF can be expressed as:

$$CF = \frac{\max\,(a_w(t))}{\text{r.m.s.}\,(a_w)}$$

where a_w is the frequency-weighted acceleration. Some vibration meters express CF in decibels (dB). To convert between the linear (standard) expression of CF and the dB notation, the following formula applies:

$$CF_{dB} = 20 \log(CF_{linear})$$

where CF_{dB} is the crest factor expressed in dB and CF_{linear} is the crest factor expressed linearly.

5.8.2.4 Vibration Dose Value (VDV)

The vibration dose value is a quantity that is only applied to whole-body vibration measurements. It was developed in response to experimental research that showed a 4th-power relationship between vibration magnitude and discomfort (see Subsection 2.3.3). Therefore, it emphasizes shocks more than the r.m.s. Another difference between r.m.s. and VDV is that VDV will always accumulate and does not decay during periods of low, or zero, vibration magnitude. Furthermore, exposure to continuous vibration will cause VDV to continuously increase, whereas r.m.s. will remain constant. These features are illustrated in Figure 5.11 and Figure 5.12. For the random vibration, VDV continuously increases to a final value of 1 m/s$^{1.75}$. For the random vibration with additional shocks, VDV shows step increases at the time of the shocks, and these do not decay like r.m.s. The final VDV for the vibration with shocks is 1.26 m/s$^{1.75}$. The relatively rapid convergence in values observed for r.m.s. does not occur for VDV. The units of VDV are m/s$^{1.75}$, although occasionally, values are expressed with "VDV" used as the unit.

Mathematically, VDV can be expressed as:

$$VDV = \sqrt[4]{\int_0^T a_w^4(t)dt}$$

where T is the measurement duration and $a_w(t)$ is the frequency weighted acceleration at time t. The structure of the defining equations for VDV and r.m.s. are similar except for the power of the exponents and the inclusion of a division by the measurement duration for the r.m.s. Due to these similarities, it is possible to estimate VDV from r.m.s. using:

$$eVDV = 1.4 a_{w\,r.m.s.} T^{1/4}$$

where $eVDV$ is the estimated VDV and T is the total exposure time measured in seconds. This estimate is only valid for signals with low crest factors (i.e., < 6).

5.8.2.5 Combining Axes

For many vibration assessments, individual measurements made in orthogonal axes should be combined. Although it can be argued that this is the "correct" approach, it is unfortunate that many standards advocate the use of the "worst axis" whereby two of the three vibration measurements remain unused. To verify which axis is the worst, all axes require measurement anyway!

For r.m.s. measurements, orthogonal axes are combined using the root sum of squares (this technique is occasionally referred to as vector sum). This can be expressed as:

$$a_{xyz} = \sqrt{a_{wx}^2 + a_{wy}^2 + a_{wz}^2}$$

where a_{xyz} is the frequency-weighted root sum of squares, and a_{wx}, a_{wy}, and a_{wz} are the frequency weighted r.m.s. accelerations in the x-axis, y-axis, and z-axis, respectively. If scaling factors are to be used (as required for some whole-body vibration applications), then the expression becomes:

$$a_{xyz} = \sqrt{k_x^2 a_{wx}^2 + k_y^2 a_{wy}^2 + k_z^2 a_{wz}^2}$$

where k_x, k_y, and k_z are scaling factors in the x-axis, y-axis, and z-axis, respectively. For VDVs, orthogonal axes are combined using:

$$VDV_{xyz} = \sqrt[4]{VDV_x^4 + VDV_y^4 + VDV_z^4}$$

where VDV_{xyz} is the combined VDV, and VDV_x, VDV_y, and VDV_z are the VDVs in the x-axis, y-axis, and z-axis, respectively. If scaling factors are to be used, then the expression becomes:

$$VDV_{xyz} = \sqrt[4]{k_x^4 VDV_x^4 + k_y^4 VDV_y^4 + k_z^4 VDV_z^4}$$

5.8.2.6 Calculation of Daily Exposures

Many jobs require a range of tasks to be performed, each with a different magnitude of vibration exposure. For hand-transmitted vibration exposures, this might be due to the worker using a variety of tools during the day; for whole-body vibration exposures this might be due to driving over different road surfaces, driving different vehicles, or performing different tasks in the same vehicle. In these cases there is no one simple measure that can be taken to be representative of the whole job. One strategy is to measure the vibration exposure for the whole of the working shift. Although this is possible, it is impractical and usually expensive. Usually, the preferred alternative is to measure the vibration exposure for each task within the job and to sum the exposures. The total daily vibration exposure for r.m.s. measurements is often termed the $A(8)$. This quantity normalizes the daily exposure to an equivalent continuous 8-h exposure level. The VDV is inherently a dose measure and so there is no need to normalize to a reference vibration duration.

For r.m.s. measurements, vibration exposures can be summed using the expression:

$$A(8) = \sqrt{\frac{1}{8} \sum_{n=1}^{n=N} a_{wn}^2 t_n}$$

where a_{wn} and t_n are the frequency-weighted r.m.s. acceleration and exposure time (in hours) for task n, and N is the number of tasks. If the individual being assessed works for more than 8 h per day, then the sum of the individual exposures could be greater than 8. Nevertheless, the 8-h reference exposure time is retained.

To extrapolate VDV measurements to a daily dose, a two-stage process is required. Initially, the total dose for each task must be established. This is calculated using the expression:

$$VDV_n = \sqrt[4]{\frac{t_n}{t_{n\,measured}} \times VDV_{n\,measured}^4}$$

where VDV_n is the total VDV for task n, t_n is the total period of vibration exposure for task n, $t_{n\,measured}$ is the time that the VDV for task n was measured, and $VDV_{n\,measured}$ is the measured VDV for task n. The second part of the process requires the VDVs from each task to be summed using the expression:

$$VDV_{total} = \sqrt[4]{\sum_{n=1}^{n=N} VDV_n^4}$$

where VDV_{total} is the daily VDV, VDV_n is the VDV for task n, and N is the number of tasks.

To calculate a daily exposure, the total daily exposure time is required. Many workers overestimate tool use times (Palmer et al., 1999). For example, in many situations, a substantial proportion of the working time is taken up with positioning the tool and workpiece. The "trigger-on" time (i.e., the true duration of vibration exposure) can be substantially lower than the reported tool-use time. This must be taken into account either by including the "trigger-off" times in the vibration measurement period and measuring full cycle times or by performing calculations according to the trigger-on time.

5.8.2.7 Acceptable Exposure-Time Thresholds

Once the vibration emission of a tool, machine, or vehicle is known, the maximum acceptable exposure time to comply with a standard, guidance, or regulation can be calculated. For r.m.s. measures, the time to reach an $A(8)$ threshold can be calculated using:

$$T_{A(8)} = 8 \times \left[\frac{A(8)}{a_w} \right]^2$$

where $T_{A(8)}$ is the time taken to reach the $A(8)$ threshold (in hours); $A(8)$ is the threshold value in the standard, guidance, or regulation; and a_w is the frequency weighted r.m.s. acceleration measured on the tool, machine, or vehicle.

For VDV, the time taken to reach a threshold value can be calculated using:

$$T_{VDV} = T \times \left[\frac{VDV_{threshold}}{VDV_{measured}} \right]^4$$

where T_{VDV} is the time taken to reach the VDV threshold (in hours), T is the duration of measurement in hours, $VDV_{threshold}$ is the threshold value in the standard, guidance, or regulation, and $VDV_{measured}$ is the VDV measured on the machine or vehicle.

5.8.3 SPECTRAL ANALYSIS

Spectral analysis is the process by which signals are converted from the time domain (i.e., the waveform, with time on the x-axis) to the frequency domain (with frequency on the x-axis). If there is energy present at any frequency, then this will appear in the spectral plot. For sinusoidal vibration, all energy occurs at a single frequency; therefore, the spectrum will contain one peak at the frequency of the sinusoid. True random vibration contains energy at all frequencies and so the spectrum will be broad. Occupational vibration sources generally have elements of both of these idealized options: there is some energy present at all frequencies, but concentrated in critical areas, usually attributable to a mechanical cause (e.g., tool speed, engine speed, road profile).

There are a multitude of signal-processing techniques available to convert signals in the time domain into the frequency domain. The most useful for human vibration assessment is the power spectral density (PSD), which is calculated from the common building block for many spectral techniques, the fast Fourier transform (FFT). Octave- and one-third octave-band analyses are occasionally used, but these methods fail to provide a detailed profile of vibration exposure.

Spectral analysis is usually carried out on unweighted vibration signals.

5.8.3.1 Fast Fourier Transforms (FFTs)

Jean Fourier was a French mathematician who lived through the French Revolution. His technique of distilling a function (e.g., a vibration waveform) into a series of component parts (e.g., vibration energy at each frequency) has proved invaluable for vibration analysis and a host of other applications. The method was developed by a series of mathematicians into the fast Fourier transform (FFT), which is capable of transforming a signal in the time domain into the frequency domain.

One problem with using FFT for human vibration analyses is that the greater the duration of the measurement that is to be analyzed (i.e., the greater the number of samples in the signal), the narrower the frequency resolution. For example, if 10 min of vibration data are to be analyzed, the FFT will generate 600 data points per Hz, with each data point representing the vibration occurring within a 1/600-Hz frequency band. Very rapidly, the FFT becomes too complex to interpret. Considering that vibration measurements for assessing human environments often last several minutes, and can last many hours, the FFT in itself is not an ideal technique to use for frequency analysis.

5.8.3.2   Power Spectral Densities (PSDs)

Power spectral density is the most common technique for analyzing the frequency content of signals for human vibration applications as it is ideally suited to the analysis of random signal types. It generates a measure of the energy contained within a frequency band. PSD splits up the original signal into shorter segments and calculates the FFT for each section. The length of each individual segment is selected such that the FFT generates an appropriate frequency resolution. For example, if a frequency resolution of 0.25 Hz is required, each segment must last 4 s. Usually, the segments overlap and are "windowed" to ensure data integrity [these processes are beyond the scope of this section; see Hammond (1998) for a more detailed approach]. The units of a PSD for an acceleration signal are $(m/s^2)^2/Hz$.

5.8.3.3   Frequency Response (Transfer) Functions

Frequency response functions were encountered in Chapter 2 when considering the transmissibility of seats and biomechanical responses of people to whole-body vibration. All mechanical systems exhibit some dynamic response characteristics and will resonate at some frequency. The response characteristics of a system at any frequency are calculated using frequency response (or transfer) functions. If the input to a system at any frequency is identical to the output at that frequency, then the frequency response function is unity with zero phase shift. It is possible for the system to attenuate or amplify the signal or to introduce some element of phase lag (time delay).

To determine the frequency response function of a system, simultaneous measurements are required at the driving point (input: e.g., a car seat or a tool handle) and the point of interest (output: e.g., the head or the shoulder). If, at any frequency, the magnitude of the input and output are identical, then the transfer function is unity. If the system amplifies the vibration, then the transfer function will be greater than unity; if the system attenuates the vibration, then the transfer function will be less than unity. The ratio of output vibration to input vibration at any frequency provides the frequency response function. PSD provides a measure of the vibration energy at each frequency. Therefore, division of the PSD of the acceleration measured at the output by the PSD of the acceleration measured at the input will generate a transfer function for all frequencies simultaneously:

$$\text{(PSD) transfer function } (f) = \sqrt{\frac{PSD_{output}(f)}{PSD_{input}(f)}}$$

This method (known as the PSD method) is accurate if vibration energy is present at the input for all frequencies, there is no background noise (e.g., from the measuring instrumentation) in the acquired signals, and the system is linear. A preferred method that is more reliable uses cross-spectral density (CSD) functions and is therefore known as the CSD method. CSD measures the relationship between two signals and also generates the phase difference between them. Therefore, a transfer function calculated using the CSD method only includes elements of the vibration signals measured at the two locations that are correlated to one another (thereby reducing the influence of noise) and also produces the phase response of the system. For a linear system with no noise in the measurement, the modulus of the transfer function obtained with the CSD method will be identical to the transfer function obtained using the PSD method. If the modulus of the transfer function calculated using the CSD method is less than that calculated using the PSD method, this indicates that there is noise in the measurements or that the system is nonlinear. The modulus of the transfer function calculated using the PSD always exceeds that calculated using the CSD method. The transfer function using the CSD method is calculated using:

$$\text{(CSD) transfer function } (f) = \frac{CSD_{input.output}(f)}{PSD_{input}(f)}$$

Figure 5.13 shows the transmissibility measured using the PSD method and the transmissibility measured using the CSD method. The two methods give similar results across most of the frequency range. However, in the resonance region, there is a discrepancy between the transmissibilities measured using the two techniques. This indicates that there is either a low signal-to-noise ratio at these frequencies or that the response of the seat is nonlinear (see also Subsection 5.8.3.4).

The usual representation for a transfer function includes the modulus, phase, and coherence.

5.8.3.4   Coherence

The coherence is the extent of correlation between an input and an output signal. If the vibration at the output is perfectly correlated to the vibration at the input, then the coherence has a value of 1. Any nonlinearities or errors in the signals (e.g., electrical noise in the data-acquisition system or interference) will reduce the coherence. In partnership with a transfer function calculated using the CSD method, coherence is a powerful tool for providing an indication of the reliability of a measurement. In the laboratory, near-perfect coherencies are possible for human vibration applications (i.e., > 0.95 at all frequencies of interest). For field measurements, there are often zones of the frequency range where the coherence drops, often

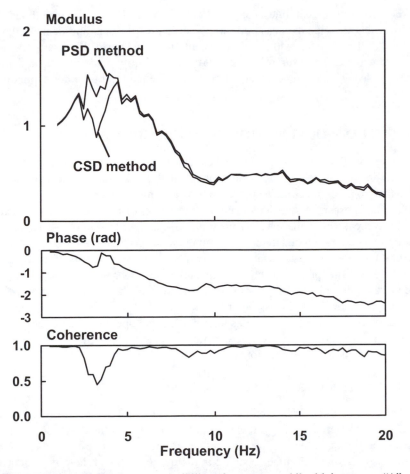

FIGURE 5.13 Modulus of the transmissibility of a car seat while driving on an "A" road calculated using the PSD method, and the modulus and phase of the transmissibility and the coherence calculated using the CSD method. The PSD method only generates the modulus of the transfer function, whereas the CSD method also generates the phase. The drop in coherence at 3 to 4 Hz corresponds to those frequencies where there is a discrepancy between the transmissibilities measured using the PSD and CSD methods.

due to lack of vibration energy at that frequency. One should be cautious in the interpretation of transfer function results in regions of low coherence.

Coherence is calculated using the CSD and PSDs of the input and output signals:

$$\text{coherence function } (f)^2 = \frac{\left| CSD_{input.output}(f) \right|^2}{PSD_{input}(f) \times PSD_{output}(f)}$$

Considering the example car-seat transmissibility shown in Figure 5.13, the zone of low coherence corresponds to those frequencies where there is a difference between the transmissibilities measured using the CSD and PSD methods. This indicates that the signals measured on the surface of the seat and beneath the seat are less well correlated within this range of frequencies. In this example, and in common with most measurements of seat transmissibility made in cars on the road, there was a low magnitude of vibration in the frequency range corresponding to low coherence.

5.9   SUCCESSFUL VIBRATION MEASUREMENT

There are many ways in which the investigator can make errors during vibration measurement. As there is always an element of the unknown in every assessment, and as most assessments take place on site in time-pressured situations, careful preparation is essential in order to minimize the risks of making mistakes. It can be helpful to break down the measurement and assessment task into a series of steps, although no generic procedure can accommodate all eventualities. Despite the focus of vibration measurements usually being related to the site visit itself, many important stages of the process occur before and after the actual measurement. The procedure described here only covers quantification of the vibration exposure magnitudes for field measurements. Other activities might be required, depending on the purpose of the measurement, such as interviewing the operator and possibly administering questionnaires (e.g., Nordic questionnaire, symptom checklists), or laboratory testing to determine emission values.

5.9.1   STEP 1: STRATEGIC PLANNING OF APPROACH

The first step of the overall vibration measurement and assessment procedure is to confirm an appropriate strategy. It is wise to consider this at the outset of the project. If tendering for a contract, the assessment technique, number of channels measured, standards applied, etc., can be specified in the quote such that there is no scope for misunderstanding of the requirements between the client and investigator.

An assessment might be for a motion sickness, hand-transmitted vibration, or whole-body vibration application, or a combination of more than one of these. The frequency range of the vibration to be assessed depends on the application area, and this often dictates the type of accelerometer and hardware to be used. It is essential that the correct frequency weightings are available. The assessment might be a simple check to establish compliance with a guidance or regulation, in which case a simple human vibration meter might be adequate, or the assessment might need more in-depth analysis whereby a data-acquisition system will be required.

At this strategic stage of the measurement process, it is useful to clearly identify which methodology in which version of which standards are to be used for the assessment. For example, the client might require a particular method due to the geographical location of his or her ultimate market where different standards are used. Some standards specify a different approach to health from one emphasizing comfort. This should be accommodated for. For some types of measurement, it is possible to use a multitude of measurement locations (e.g., assessment of whole-body vibration exposures for bus passengers); a pragmatic but valid decision of how many locations to include should be made at this early stage so that an appropriate number of measurement channels are supplied.

Essentially, a good strategic approach to the measurement process will reduce the risk of being unable to complete the analysis during the final phases of the assessment (when it may well be too late).

5.9.2 Step 2: Collation and Calibration of Equipment

Once the appropriate measurement strategy has been decided, the required measuring instrumentation will need collation and calibration. A preliminary calibration can be carried out (as described in Section 5.7) at the office or laboratory. This approach has the advantage of giving the investigator confidence that hardware (and software) settings are nominally correct, thereby reducing time on-site for configuring equipment. It also has the practical benefit of functionally testing all equipment prior to leaving for the site visit. Calibration should also be checked immediately prior to the measurement itself on site, where minor adjustments might be necessary.

5.9.3 Step 3: Mounting of Accelerometers

For whole-body vibration field measurements, accelerometers should be mounted on the seat surface and possibly on the floor (beneath the seat or at the feet) and at the backrest. If measures are to be made on the surface of a compliant seat, then seat accelerometers should be mounted in a flexible disc (Figure 5.14). This disc is sometimes referred to as an "SAE pad" due to its first appearance in a Society for Automotive Engineers standard (SAE, 1973), although it is now also defined elsewhere (e.g., ISO 10326-1, 1992). The disc has a diameter of 25 cm and the accelerometer is mounted onto a 7.5-cm-diameter thin metal disc in the center of the device. The mounting device is placed onto the center of the seat cushion (Figure 5.15). When the seat occupant sits on the disc, his or her body weight presses the metal part of the disc onto the surface of the seat so that the accelerometer measures the vibration in the seat center. The flexible part of the disc deforms to the contour of the occupied seat and allows for the accelerometer to be sat on without causing excessive discomfort. The same pad is also sometimes used for measurement of backrest vibration, where it is fixed to the seat back. Many discs contain a triaxial

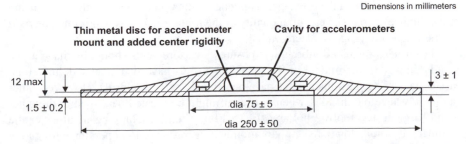

FIGURE 5.14 Design of flexible disc for mounting seat accelerometers as defined in ISO 10326-1 (1992).

FIGURE 5.15 Flexible disc containing accelerometers mounted on the seat of a forestry machine.

accelerometer set, in which case the disc must be orientated such that the x-accelerometer is aligned in the x-direction.

To measure seat transmissibility, accelerometers must also be mounted beneath the seat. Ideally, these should be mounted on the floor directly beneath the seat pad on which the seat occupant sits. However, for many vehicles (e.g., cars) there are practical difficulties with mounting on the floor beneath the seat. Therefore, accelerometers are sometimes mounted on seat guides and at seat mounting points. Ideally, measurements should be made at many points and mathematical techniques used to calculate the vibration beneath the center of the seat. However, if it can be assumed that the floor is rigid and that there is negligible pitch and roll vibration, then a mounting point on one seat guide will suffice. Many vehicles have trim and covers that must be removed to gain access beneath the seat; this is not always straightforward! Floor accelerometers can be mounted using cyanoacrylate adhesive (superglue), double-sided adhesive tape, magnetic fixings, beeswax, or clamping using hose (jubilee) clips. The essential quality of the fixing method is that it must be rigid in the frequency range of interest.

For hand-transmitted vibration exposures, measurements should be made at the hands, ideally in the middle of the grip. Some tools have two clearly defined (and used) handles (e.g., a pneumatic pavement breaker or rock drill). However, other tools only have one handle, where a trigger might be located, and require the other hand to guide the tool by holding the tool body (e.g., a chipping hammer or die grinder). In some situations, the vibration is transmitted to the operator through the workpiece and not through the tool itself (e.g., pedestal grinding). The general principle of measurement at the center of the hand contact surface should be applied

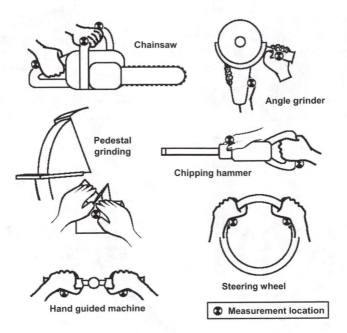

FIGURE 5.16 Examples of practical measurement locations for some common power tool types (from BS EN ISO 5349-2, 2002).

wherever possible, although practical difficulties might require a compromise of mounting accelerometers adjacent to the contact point. Suggested measurement locations are published in ISO 8662-2 to ISO 8662-14 and ISO 5349-2 (2001; Figure 5.16). Care must be taken to ensure that transducers, mountings, and their associated cables do not compromise the safety of the tool operation.

The physical mounting of the accelerometers for hand-transmitted vibration applications can be achieved by a variety of methods. Essentially, the accelerometers can either be mounted to the tool (or workpiece) or they can be mounted onto handheld adapters that are located between the operator's hands and tool. Some methods of fixing to the tool can cause minor damage to the tool handle but provide a convenient mounting point. These include using a threaded mounting stud screwed into the tool and accelerometer or using cement (possibly in combination with a disposable mounting stud; Figure 5.17). An alternative, but more bulky, method of fixing accelerometers to the tool is to use mounting blocks clamped in place using hose clips (Figure 5.18A). Handheld adapters can be used in situations where fixing to the tool is inappropriate (e.g., the handle of the tool might be covered in a compliant hand-grip). The adapters either contain an accelerometer or an accelerometer mounting point and are pushed onto the tool by the gripping force of the operator (Figure 5.18B). The "palm adapter" used for measurement of the dynamic properties of antivibration gloves can be used as a handheld adapter. Some advantages and disadvantages of the alternative accelerometer mounting methods for hand-transmitted vibration are summarized in Table 5.2.

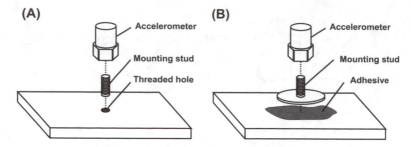

FIGURE 5.17 Methods of mounting an accelerometer to a tool. Accelerometers can be mounted using a stud that is screwed into a threaded hole (A) or adhered to the tool using glue or wax (B).

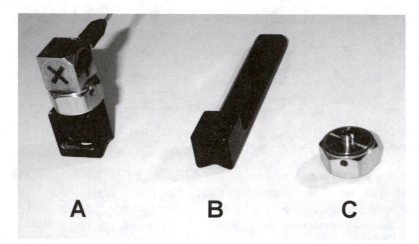

FIGURE 5.18 Methods of clamping an accelerometer to a tool. Accelerometers can be clamped using a mounting block that is fastened to the tool using a hose clip (A) or held on the tool by the gripping force using a hand-adapter (B). It is often necessary to use a mechanical filter if there is a risk of DC shifts occurring (C).

Sometimes high-frequency high-magnitude impulsive vibration can affect the materials from which the accelerometer is constructed. They are a particular problem for measurements of vibration on impulsive and percussive tools (e.g., chipping hammers). These phenomena are known as "DC shifts" and can be observed in the time domain either as instantaneous changes in the offset of the accelerometer or as an exponential decay, due to the functioning of band-pass filters in the signal conditioning (Figure 5.19). If DC shifts occur, then a low-frequency artifact will occur in the measurement and this will dominate the values obtained in the vibration assessment, especially when the signal is W_h weighted. To reduce the occurrence of DC shifts, a mechanical filter should be mounted between the accelerometer and the mounting point (Figure 5.18C). Mechanical filters are vibration isolation devices that transmit the low frequencies of interest to the accelerometer but protect it from the shocks. Although it might seem peculiar to mount an accelerometer on a vibration

TABLE 5.2
Summary of Advantages and Disadvantages of Alternative Accelerometer Mounting Methods for Measurement of Hand-Transmitted Vibration

Mounting Type	Advantages	Disadvantages
Stud mounting	Good frequency response Not affected by surface temperature	Contact surface must be flat Cannot be used on hand tools where it might affect the electrical or pneumatic safety of the power tool Damages surface
Glue	Good frequency response	Contact surface must be flat and clean Damages surface
Cement/epoxy resin	Good frequency response Fits to uneven surfaces	Contact surface must be clean Damages surface
Metal "U" clamp	Suitable for triaxial measurements	Bulky and heavy
Mounting block with metal hose-clip	Suitable for triaxial measurements Rapid mounting Relatively light No sharp edges	Mainly limited to measurement on power tool handles
Simple handheld adapter	Can be used in cases where a fixed coupling is inapplicable, e.g., on soft or resilient materials Rapid mounting	Only suitable for fixed-hand position and where the handle is always being held Frequency response depends on surface material The presence of the adapter may change the operation of the power tool and the resulting vibration magnitude Additional fixing (e.g., adhesive) is required for transverse vibration measurements
Palm adapter	Suitable for use inside gloves Measurement includes dynamic properties of glove Rapid mounting	Difficult to monitor alignment of adapter Difficult to use for triaxial measurements Requires miniature accelerometers that can be expensive and relatively fragile
Individually moulded adapter	Can be used in cases where a fixed coupling is inapplicable, e.g., on soft or resilient materials Little influence of the adapter on the operation of the power tool Fair frequency response	Preparation of the adapter is a laborious, time-consuming procedure Difficult to use for triaxial measurements

Source: Adapted from International Organization for Standardization (2001). Mechanical vibration: measurement and evaluation of human exposure to hand transmitted vibration — part 2: practical guidance for measurement at the workplace. ISO 5349-2. Geneva: International Organization for Standardization.

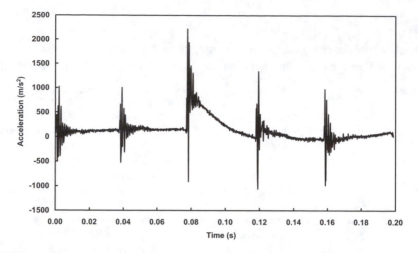

FIGURE 5.19 Acceleration measured on the handle of a pick hammer showing a DC shift at about 0.08 s followed by an exponential decay due to the characteristics of the filter in the signal conditioning.

isolation mount, the device should only filter those frequencies of vibration higher than the low-pass cutoff frequency of the antialiasing filter.

5.9.4 Step 4: Measurement of the Vibration

Once all equipment is calibrated, mounted, and connected, the measurement itself can be instigated. It is worth checking that all settings and configurations are correct (these might be stored in the measuring instrumentation and might not be reloaded by default). This check should include verifying that the appropriate frequency weighting has been selected and that antialiasing filters are set correctly, as these might have been switched off during the calibration process.

For human vibration meters, the internal memory buffer should be reset so that previous measurements are not inadvertently included in the new analysis. During the measurement, the meter's display should be monitored to ensure a good signal level (usually indicated by a bar-type display) and that no overloads occur. Gain settings might need adjustment if the signal level is too high or too low. After each measurement, the results should be noted and stored in the meter.

For data-acquisition systems, the signal should be checked to ensure that the full range of the measurement hardware is used and that the signal has not been clipped. The data should be saved within the system and notes made to ensure unique identification of each measurement. Many systems allow for descriptors to be added to each file, and these should be used.

Many subtasks can produce vibration of different characteristics. For example, a pavement breaker's vibration emission will be different while cutting tarmac or the softer ground beneath the road surface. It is therefore beneficial to perform a basic task analysis to identify stages within the job that can then be assessed separately, so that an overall understanding of the vibration exposure profile can be attained. Operators often report that one part of their job exposes them to more vibration than another; this can be a helpful indicator of which parts of the cycle might dominate.

If the cycle time is short, it might be appropriate to measure for one or more complete cycles, in addition to short measures of each subtask (e.g., a pedestal grinding task). For longer exposures, a representative sample is usually adequate (e.g., driving a truck). It is difficult to specify a minimum acceptable sample time for vibration measurements. For hand-transmitted vibration, samples of at least 20 s should be measured with at least three repeat measures taken. If the vibration occurs in short bursts, then more individual measures are required. A rule of thumb is that the total duration of vibration measured should be at least 60 s. Of course, more repeats and longer durations of measurement are preferable. For whole-body vibration, longer measures are required to ensure reliable measurements at low frequency. If possible, each measurement should last at least 3 min with at least three repeat measures taken. If testing is carried out on public highways, then maintenance of road safety must be stressed to the driver, as they can occasionally become preoccupied with maintaining a steady fixed speed for the purposes of the assessment to the detriment of their situational awareness. There are very few urban streets where an uninterrupted measurement of 3 min can be completed.

Considering the relatively short amount of time that is required for each individual measurement in comparison with the setup time and, usually, travel to site, it is appropriate to make many measurements to check repeatability (poor repeatability does not necessarily indicate a problem, but it should be reported nevertheless so that a worst-case scenario can be modeled).

Care should be taken throughout the measurement process to avoid artifacts in the signal that are not vibration exposures. For example, sitting on or vacating an instrumented seat can generate large shock signals, as can the process of picking up or placing down a hand tool. These events can dominate the vibration measurement. However, other elements of the exposure that could be assessed include starting, running up to speed, and stopping tools. For most measurements, gaps in the vibration exposure should be avoided, as these will induce a decay in the r.m.s. measurement. Therefore, the meter or acquisition should be started after the vibration has commenced and stopped before the vibration ceases.

5.9.5 Step 5: Analysis and Postprocessing

The appropriate standards, guidelines, and regulations for vibration assessment should have been identified in Step 1 of the measurement process. It should therefore be a fairly straightforward procedure to calculate overall exposure levels and acceptable exposure times by using the techniques described earlier in this chapter.

Comparison of the exposures to threshold values in standards is essential, but simply identifying noncompliance without identifying possible target areas for a risk reduction strategy limits the value of the assessment. Therefore, breaking the operation down into subtasks (as suggested in Step 4 of the measurement process) can identify those parts of the operation that require attention. Frequency analysis of these subtasks will allow for recommendations to be made regarding which frequencies of vibration should be reduced. This information can be used in the selection of personal protective equipment (e.g., replacement seats). Occasionally, presentation of time histories can be helpful in illustrating the nature of the vibration (e.g., showing the dominance of shocks in the signal). For other applications, transfer functions require calculation to identify the dynamic performance of mechanical elements in the vibration transmission path.

Even a relatively straightforward vibration assessment can generate an enormous volume of data. The report should therefore be logically and simply presented, so that a lay reader can interpret the key conclusions. Usually, this will focus on the allowable exposure times and opportunities for risk reduction. ISO 5349 suggests that for hand-transmitted vibration assessments, the report should include up to nine categories of information, including descriptions of the tools tested, tasks selected, and equipment used for the measurement, in addition to the vibration magnitudes themselves (Table 5.3). These principles can also be useful for the selection of information required in a whole-body vibration assessment report.

5.10 CASE STUDY 1: HAND-TRANSMITTED VIBRATION ASSESSMENT

This section provides an example of a hand-transmitted vibration assessment. The duration required for a small hammer drill to exceed the Physical Agents (Vibration) Directive action and limit values was to be determined. The drill was electrically powered and contained a 550 W motor. It had two handles and was typically used to drill into brick.

As only a simple hand-transmitted vibration assessment was required, the only frequency weighting required was W_h, and the assessor had the option of choosing to use either a vibration meter or data-acquisition system. The Directive to which the measurement was to be made requires a triaxial measurement at the hands. As both hands contact with the tool, measurements were required at two locations. Therefore, ideally, two triaxial accelerometers should have been available so that measurements could be made at both hands simultaneously. However, according to the required measurement standard, it would also be acceptable to use a simpler setup such that each axis for each hand was assessed separately. (This would therefore require six times as many measurements as for the example where all axes for both hands were measured simultaneously.)

For this case study, two human vibration meters were available, each of which could measure at least three axes of vibration simultaneously. Prior to leaving for the site visit, the meters and accelerometers were calibrated and set up such that no programming would be required later. Furthermore, a full range of attachments and handheld adapters were collated to ensure that the most appropriate accelerometer mounting method could be used. The calibration was checked on site prior to mounting of the accelerometers; this required changing the setup of the meters such that the frequency weightings were bypassed.

TABLE 5.3
Information to Be Reported, Depending on the Situation Investigated, Following a Human Vibration Assessment

Category	Information to be Reported
General information	Company/customer
	Purpose of the measurements (e.g., evaluation of vibration exposure of individual workers, worker groups, evaluation of control measures, epidemiological study)
	Standard, guidance, or regulation to which assessment was made
	Date of evaluation
	Subject or subjects of the individual exposure evaluation
	Person carrying out the measurements and evaluation
Environmental conditions at the workplace	Location of measurements (e.g., indoor, outdoor, factory area)
	Temperature
	Humidity
	Noise
Information used to select the operations measured	Results of interviews with workers
	Results of simple task analysis
Daily work patterns for each operation evaluated	Description of operations measured
	Machines and inserted tools used
	Materials or workpieces used
	Patterns of exposure (e.g., working hours, break periods)
	Information used to determine daily exposure times (e.g., work rate or numbers of work cycles or components per day, durations of exposure per cycle, or handheld workpiece)
Details of vibration sources	Technical description of the power tool or machine
	Type and/or model number
	Age and maintenance condition of the power tool or machine
	Weight of the handheld power tool or handheld workpiece
	Vibration control measures on the machine or power tool, if any
	Type of handgrip used
	Automatic control systems of the machine (e.g., torque control on nut runners)
	Power of the machine
	Rotational frequency or percussive speed
	Models and types of inserted tools
	Any additional information (e.g., unbalance of inserted tools)
Instrumentation	Instrumentation detail
	Calibration traceability
	Date of most recent verification test
	Results of functionality check
	Results of any interference tests

TABLE 5.3 *(Continued)*
**Information to Be Reported, Depending on the Situation Investigated,
Following a Human Vibration Assessment**

Category	Information to be Reported
Acceleration measurement conditions	Accelerometer locations and orientations (including a sketch and dimensions)
	Methods of attaching transducers
	Mass of the transducers and mount
	Operating conditions
	Arm posture and hand positions (including whether the operator is left- or right-handed)
	Any additional information (e.g., data on feed and grip forces)
Measurement results	x-, y-, and z-axis frequency weighted hand-transmitted vibration values
	Measurement durations
	Unweighted frequency spectra
	If single- or two-axis measurements were used, the multiplying factors to give vibration total value estimates (including justification for using single- or two-axis measurements and justification for the multiplying factors used)
Daily vibration exposure evaluation results	Vibration total values for each operation
	Duration of vibration exposure for each operation
	Partial vibration exposures for each operation (e.g., subtasks)
	Daily vibration exposure, $A(8)$
	Evaluation of the uncertainty of daily vibration exposure results
	Allowable exposure times to reach threshold values

Source: Adapted from International Organization for Standardization (2001). Mechanical vibration: measurement and evaluation of human exposure to hand transmitted vibration — part 2: practical guidance for measurement at the workplace. ISO 5349-2. Geneva: International Organization for Standardization.

As the routing of the internal electrical wiring of the drill was unknown, it was not appropriate to tap a hole for mounting of a stud. It was also desirable that the tool would not be damaged by the testing, and so adhesive methods of accelerometer fixing were not used. As the handles were rigid, the assessor was able to use mounting blocks or handheld adapters to fix the accelerometers to the tool handles. Mounting blocks with metal hose clip clamps were selected as the preferred mounting method due to their superior frequency response.

Prior to the vibration measurement itself, the settings of the meters were inspected to ensure that the frequency weighting was reapplied following the calibration check. The meters were reset to clear the memory buffer. The tool operator was instructed to drill one hole in a typical section of the brick wall. This task took about 60 s to complete. Immediately after the drilling task commenced, the meters were started and the vibration was measured for 30 s. During this first test, one axis of one of the meters overloaded, and so the data from this test were discarded and the gain was reduced for that channel. The measurement process was repeated so

TABLE 5.4
Results from Example Vibration Assessment of an Electrical Hammer Drill

Measurement	Frequency-Weighted Vibration Magnitude (m/s² r.m.s.), Left Hand				Frequency-Weighted Vibration Magnitude (m/s² r.m.s.), Right Hand			
	x-axis	y-axis	z-axis	r.s.s.	x-axis	y-axis	z-axis	r.s.s.
1	16.1	8.1	7.8	19.6	13.8	7.9	8.2	17.9
2	13.8	7.2	6.9	17.0	11.1	7.4	7.3	15.2
3	15.3	7.9	7.5	18.8	12.8	7.4	7.5	16.6
4	17.7	9.0	8.8	21.7	13.2	8.0	8.5	17.6
5	15.1	7.3	6.8	18.1	12.9	6.5	7.0	16.1
6	14.9	7.2	7.1	18.0	13.0	6.7	6.9	16.2
			Mean r.s.s. vibration	**18.9**			Mean r.s.s. vibration	**16.6**

that six measures were made, with no overloads. Results were stored in the meter and later transferred to a computer using a cable link.

The results from this example vibration assessment are shown in Table 5.4. These show that for the left hand, the vibration was dominated by x-direction vibration, and that for the right hand, the vibration was also dominated by x-direction vibration. As these measurements were to be used to provide guidance for acceptable exposure times, individual axes for each hand required combining using the root sum of squares method. These combined data showed that the magnitude of the vibration at the left hand was greater than the magnitude of the vibration at the right hand. The acceptable use times were calculated in accordance with the action value (2.5 m/s²) and limit value (5 m/s²) specified in the Physical Agents (Vibration) Directive. Therefore the tool could be used for 8 min before exceeding the exposure action value and 34 min before reaching the limit value, based on average results from the six measurements.

Although this procedure is valid, there are ways in which the assessment could have been improved. One improvement would have been to measure the vibration emission of the tool while fitted with a variety of cutting bits and drilling into a variety of materials. Each of these bit and material combinations would require enough measurements to ensure validity and the full analysis procedure. With this additional information, a worst-case, and best-case, situation could be found; alternatively, different limits could be set depending on the type of work being carried out. A further improvement would have been to use more than one operator as this could improve confidence in applying the results generally.

If a data-acquisition system was used in preference to the vibration meters, then additional information could have been presented. For example, an example of the time-domain signal could have been included in the report to illustrate the impulsive nature of the vibration exposure. Also, the frequency content of the signal could have been calculated.

5.11 CASE STUDY 2: WHOLE-BODY VIBRATION ASSESSMENT

This section provides an example of a whole-body vibration assessment. The purpose of the assessment was twofold: first, to determine whether the driver of a heavy goods vehicle (HGV) would exceed the action and limit values from the Physical Agents (Vibration) Directive during their typical 8-h shift, and second, to determine the effectiveness of their seat in providing isolation from the vibration.

For assessment of the vibration exposure according to the Physical Agents (Vibration) Directive, measurements are only required on the seat surface. However, to determine the dynamic performance of a seat, additional measurements are required at the seat base. As HGV vibration is usually dominated by the vertical component and the Directive only specifies that the "worst axis" be considered, the assessment could be completed by using two single-axis accelerometers: one mounted on the surface of the seat and one mounted beneath the seat. As only vertical whole-body vibration was to be assessed, the W_k frequency weighting was required. As the dynamic characteristics of the seat were to be measured, a data-acquisition system was necessary so that seat transmissibility could be calculated.

The system to be used for the measurements utilized two piezoresistive accelerometers (one fitted into an SAE pad), an accelerometer amplifier with integral antialiasing filters (set, for this example, at 125 Hz) and a portable computer fitted with a 16-bit ADC card. Prior to leaving for the site visit, the accelerometers were calibrated using gravity. The computer was set up such that it would acquire the raw acceleration signals (i.e., unweighted) on two channels at 512 samples per second, and this setup was saved to disk. Tools were also collated to ensure that any trim surrounding the seat base could be removed, and replaced following the measurement. All batteries (and spares) were fully charged. The calibration was checked on site prior to the mounting of the accelerometers.

The seat of the HGV was a suspension seat and the trim surrounding the suspension mechanism was easily removed. It was possible to fix the accelerometer in the center of the seat base using cyanoacrylate adhesive. Cables were routed to ensure that they would not interfere with the moving parts of the seat. The SAE pad was placed in the center of the surface of the seat.

Prior to vibration measurement itself, the setup of the data-acquisition system was checked to ensure an appropriate sample rate and frequency for the antialiasing filters. A circuit was driven that included typical roads on which the vehicle traveled on a daily basis. These included a motorway section, A-road, and roads close to the HGV depot on an industrial park. For the motorway and A-road, three 180-s measurements were made while traveling at a steady speed. For the roads on the industrial park, nine 60-s measurements were made, as it was not possible to achieve the longer, ideal, uninterrupted measurement durations. The driver was not aware of the precise start and stop times for the measurements so that they would not be distracted from their primary (i.e., driving) task. Each measurement was given a distinctive file name and use was made of the text descriptions available within the software. The driver was interviewed to obtain an estimate of the times driving on different road types during a typical day.

Postprocessing and analysis involved frequency weighting the signals and calculating the frequency-weighted r.m.s. and VDV on the seat and at the base of the seat. From the driver's estimate of exposure times on each road type and the results from the seat surface, the total daily dose could be calculated for both VDV and r.m.s. (Table 5.5). Comparison of the r.m.s. results with the Physical Agents (Vibration) Directive action value (0.5 m/s^2) and limit value (1.15 m/s^2) showed that the driver's exposure exceeded the action value but not the limit value. Similarly, the VDV results exceeded the action value (9.1 m/s$^{1.75}$) but not the limit value (21 m/s$^{1.75}$).

The SEAT value was calculated using both the r.m.s. and VDV methods by taking the ratio of the vibration magnitude on the surface of the seat to the vibration magnitude at the base of the seat, and had a mean value of 97%, indicating that it did not provide good isolation from the vibration at the cab floor (values of about 60% can be achieved for some HGV seats). The seat transmissibility was calculated using the CSD method and showed a peak in the dynamic response of the seat at about 2.5 Hz (e.g., Figure 5.20). The transmissibility was calculated for each measurement and the median taken for each road type.

There are areas where this measurement could have been expanded. First, more axes of vibration could have been measured on the surface and at the base of the seat. These data could have been used to confirm that the vibration in the vertical direction was dominant [for the purposes of the assessment and for comparison with the Physical Agents (Vibration) Directive] and also to quantify the exposures in the nonvertical directions. Additionally, the dynamic performance of the seat in nonvertical axes could also have been analyzed. A second possible area of expansion would have been to make measures on a range of vehicles to investigate repeatability of the measurements between nominally identical vehicles, across a range of vehicle types within a class, and across classes. Finally, an increased number of measurements could have been made by including more roads and more repeats on each road.

5.12 CHAPTER SUMMARY

The vibration measurement task is a process of converting a mechanical quantity into a number or figure, usually presented in a report. The multistage process includes a mechanical phase (i.e., mounting of an accelerometer), an electrical phase (i.e., conversion of electrical properties into a conditioned voltage), and a digital signal-processing phase (i.e., frequency weighting and performing calculations as required to meet the objectives of the measurement). There are two types of measurement systems available: human vibration meters that are compact and self-contained, and data-acquisition systems that are modular, more powerful, and more flexible in their application but more complex (and often more expensive). For both types of systems it is essential that they are configured correctly; otherwise measurement reliability will, at best, be compromised, or, at worst, make the results unusable. Vibration meters can usually provide measurements of the statistical properties of the vibration (e.g., frequency-weighted r.m.s. accelerations, peak accelerations, crest factors,

TABLE 5.5
Results from Example Vibration Assessment of a Heavy Goods Vehicle

Road Type	Driver's Estimated Daily Exposure for Road Type	Duration of Each Measurement	Vibration Magnitude on Seat (m/s² r.m.s.)	Mean Vibration Magnitude on Seat (m/s² r.m.s.)	Vibration Dose Value on Seat for Duration (m/s$^{1.75}$)	Mean Vibration Dose Value on Seat for Duration (m/s$^{1.75}$)
Motorway	4 h	3 min	0.37	0.38	1.9	2.0
			0.35		1.8	
			0.41		2.2	
A-road	3 h	3 min	0.61	0.62	3.2	3.3
			0.67		3.7	
			0.58		3.1	
Industrial park	1 h	1 min	0.71	0.75	2.9	3.1
			0.66		2.7	
			0.80		3.4	
			0.71		2.9	
			0.73		2.8	
			0.81		3.7	
			0.75		3.2	
			0.76		3.0	
			0.79		3.2	
			Estimated daily exposure (r.m.s.)	0.54	**Estimated daily exposure (VDV)**	10.9

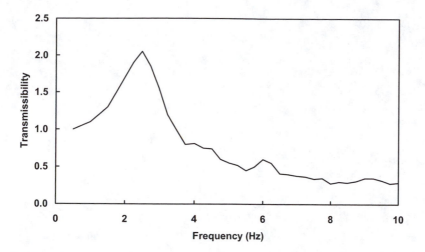

FIGURE 5.20 Modulus of the seat transmissibility measured for a suspension seat mounted in a heavy goods vehicle while driving on roads in an industrial park.

VDVs). Data-acquisition systems can be programmed to calculate all statistical measures but can also be used to generate graphs of the time histories of the vibration and to make calculations in the frequency domain.

A system for successful vibration measurement can be broken down into a five-step process: strategic planning of approach, collation and calibration of equipment, mounting of accelerometers, measurement of the vibration, and analysis and postprocessing.

6 Whole-Body Vibration Standards

6.1 INTRODUCTION

Almost every electrical and electronic item, piece of clothing, toy, or item of furniture displays text or a logo to convey that the item complies with a standard. Consumers might find some comfort in knowing that these labels are there, but are generally unable to fully interpret their meaning and would probably not notice if some were absent. The International Organization for Standardization (ISO) provides the definition of a standard:

> Standards are documented agreements containing technical specifications or other precise criteria to be used consistently as rules, guidelines, or definitions of characteristics, to ensure that materials, products, processes, and services are fit for their purpose.

For some items to be fit for their purpose, there must be a consensus regarding formatting, e.g., the magnetic strip and chip on every credit card worldwide should work with every credit card reader worldwide. For other items, safety is of paramount importance, e.g., the material dimensions and properties of toys must not pose a hazard to children. For still other items, the performance is critical, but it is difficult for the purchaser to test the item prior to purchase, one example being a suspension seat that must isolate the occupant from vibration, not amplify it. For these example applications, standards can be used to ensure a minimum compliance with a "fit for purpose" criterion.

Standards are also used to specify measurement methodologies. This is to ensure that if a set of methods is used by one laboratory, then similar results would be obtained by another laboratory using the same set. Ideally, either of these laboratories could simply report that the measurement was made "according to Standard (standard number)," and no further explanation would be needed (although further explanation might be desirable if the reader were not familiar with the standard). Within the discipline of human vibration research, one of the most useful (and yet controversial) applications for standards is in providing generalized methods for measurement.

Standards can be used at all stages of a product's life. At the concept stage, compliance with a standard can be an essential part of the requirements for the product. Standards are then used by certification bodies to check that the product meets minimum requirements prior to being released for sale. Sometimes, certification is mandated. For example, many products (including suspension seats) must

FIGURE 6.1 The "CE" mark that indicates that a product complies with a European Directive.

meet minimum legal requirements (often defined in standards) before they can be sold within the European Community and carry the "CE" mark (Figure 6.1). Potential customers for the product might also refer to standards to assist their purchasing decision. While the product is being used, standards might also be used by those testing its performance. For example, should an individual claim that an injury has been caused by the vibration emission of a tool, then standardized methods of vibration assessment can be used.

Most vibration standards seek to enable the quantification of the emission of the machine or the exposure of the operator. The terms *emission* and *exposure* both relate to the vibration experienced from the machine. There is, however, an important difference. The emission is the vibration magnitude that can be specified for the machine operating under one particular set of working conditions. If the working conditions change, then the emission will change. Vibration exposure combines the emissions of the machine with the working time and working condition profile to generate a magnitude of vibration for a specific exposure duration (often a working day). Therefore, emission is specific to the machine; exposure is specific to the operator.

Some standards have been considered elsewhere in this book. For example, some parts of standards concerned with vibration perception have been covered in Chapter 2. This chapter focuses on standards concerned with whole-body vibration; Chapter 7 is concerned with hand-transmitted vibration standards. Standards are developed by a variety of bodies (Section 6.2) and compliance can sometimes be mandated by other documents (Section 6.3). Standards exist for measuring vibration (Section 6.4), determining typical vibration exposure during work (Section 6.5), measuring emission values (Section 6.6), or for laboratory testing of seats (Section 6.7). This chapter primarily focuses on standards that have been published since 1990, but does not seek to give an exhaustive coverage and tends to focus on the most widely used samples. Many earlier publications (some of which are still current) are summarized by Griffin (1990).

6.2 STANDARDIZATION BODIES

Standards are developed by committees working at the national, regional (e.g., European), and global level. Members of the national committees are usually experts, representing companies, professional bodies, research institutes, and national health and safety bodies. Members of international committees are nominated by the national committees. These experts are not paid by the standards organizations, but work on a voluntary basis or are supported by their employers. Most standards take years to develop from their initial draft through to a final published document. The process is therefore expensive in terms of time, travel, and subsistence costs, and most of these costs are borne by the committee members' organizations. There are benefits to all of these groups if standards are developed appropriately. For example, companies can anticipate and shape future requirements for their products; professional bodies can ensure that their members are prepared for future changes; researchers can apply their knowledge to benefit society at large; health and safety bodies can ensure that appropriate levels of protection are introduced. However, the cumulative effects of getting it wrong can also be substantial, and ensuring that this does not happen can also be a motivation for an individual to participate in a committee.

As with all committees, individual members have different views and perceptions. One could argue that the accommodation of a diverse set of views is the very reason to develop broadly applicable documents (i.e., standards) in this way. Therefore, every standard is likely to contain some form of compromise in at least some area of detail. A view on whether any compromise is acceptable or not often depends on which member of the committee is consulted!

6.2.1 INTERNATIONAL ORGANIZATION FOR STANDARDIZATION (ISO)

The International Organization for Standardization (ISO) is based in Geneva, although its meetings are held worldwide. It was originally established in 1947 to harmonize the national standards that were being developed with the intention of removing barriers to international trade. Until the 1970s, ISO did not produce standards of its own. Today, standards are developed through ISO's Technical Committees that are formed from representatives of national standards bodies. Each committee is administered by a national body. Therefore, most of the work of ISO is decentralized.

Each ISO Technical Committee (TC) is referred to by a numerical code. For example, ISO TC 159 is for ergonomics; ISO TC 108 is for mechanical vibration and shock. Each TC is split into a range of subcommittees (SC). For example, ISO TC 108 has six SCs; SC 4 is for human exposure to mechanical vibration and shock (Figure 6.2). Each SC is split into a range of working groups (WG). For example, ISO TC 108 SC 4 has seven working groups; WG 2 concerns whole-body vibration.

Standards are developed through a six-stage process, from the initial proposal through to publication (Table 6.1). At the proposal stage, at least five member organizations must agree to actively participate in the work item, the work must be considered to have market relevance, and the majority of the ISO members must vote to approve the work. At this point, a number is assigned to the standard that is to be developed. The group of at least five active participants then develops the standard into a first committee draft. This is usually given a reference of CDxxxx where xxxx is the number of the standard. The standard progresses from stage to stage (usually CD to DIS to FDIS to an International Standard) as a consensus is

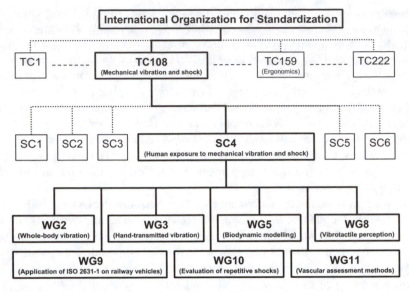

FIGURE 6.2 Organizational structure of ISO TC 108 SC 4 "Human exposure to mechanical vibration and shock." Technical committees other than TC 1, TC 108, TC 159, and TC 222 are omitted for clarity.

TABLE 6.1
Stages in the Production of an ISO Standard

Stage	Description	Deliverable	Abbreviation
1. Proposal stage	New work item proposed	New work item proposal	NP
			NWI
2. Preparatory stage	Building consensus within working group	First Committee Draft or Publicly Available Specification	CD
			PAS
3. Committee stage	Building consensus within TC or SC	Draft International Standard, Technical Specification, or Technical Report	DIS
			TS
			TR
4. Enquiry stage	Enquiry and comment on DIS	Final text for processing Final Draft International Standard	FDIS
5. Approval stage	Formal vote on FDIS	Final text of International Standard	
6. Publication stage	Publication of International Standard	International Standard	ISO

reached across those voting members of the SC (usually amounting to at least a two-thirds majority in favor). Each stage can involve a number of draft documents. Once published, each standard is reviewed at least once every 5 years to decide whether it should be revised, withdrawn, or maintained in its current form.

6.2.2 EUROPEAN COMMITTEE FOR STANDARDIZATION (CEN)

The European Committee for Standardization (CEN) is managed from Brussels, Belgium. It was initially founded in 1961 and became the only recognized cross-Europe standardization body through Directive 83/189 (1983). The remit of CEN extends beyond the European Union member states and includes countries such as Norway and Switzerland. There are similarities in the method of functioning to ISO; i.e., a decentralized committee system with meetings occurring across Europe, comprising representatives of national standardization bodies. Its aims are to support the European Single Market and to enhance the competitiveness of European representatives in the global market. CEN produces its own "EN" standards, although many (about 40%) are direct adoptions of ISO standards. All CEN standards are published in English, French, and German, although other language versions are occasionally available. All EN standards must be implemented by EU member states either by publication of a national standard with identical text (or a translation) or by endorsing the original EN document.

The need for harmonized standards between ISO and CEN resulted in the "Lisbon Agreement" (1989), and subsequently the "Vienna Agreement" (2001). These agreements acknowledge that it is often unnecessary and certainly undesirable to duplicate work within related ISO and CEN committees, especially considering that there is usually a substantial overlap between the active committee members. Therefore, standards are often given a dual reference code (e.g., EN ISO 13090-1: 1998 Mechanical vibration and shock — Guidance on safety aspects of tests and experiments with people — Part 1: Exposure to whole-body mechanical vibration and repeated shock).

Each CEN Technical Committee (TC) is referred to by a numerical code. For example, CEN TC 122 is for ergonomics; CEN TC 231 is for mechanical vibration and shock. Each TC is split into a range of working groups (WG). For example, CEN TC 122 has seven working groups; WG 1 is whole-body vibration (Figure 6.3).

Standards are developed through a seven-stage process, similar to the process undertaken for international standards (Table 6.2). For parallel development of standards between ISO and CEN, either party can take the lead whereby the work item is transferred to the appropriate body, and there is parallel voting within both organizations.

6.2.3 NATIONAL STANDARDIZATION BODIES

Historically, virtually all standards were developed by national standardization bodies. Although the standardization process at the national level still occurs, the workload involved and desire for harmonized standards regionally and globally, means that there is a greater incentive for national bodies to have a policy of adoption

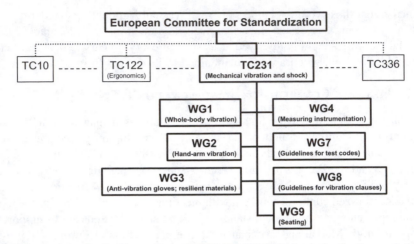

FIGURE 6.3 Organizational structure of CEN TC 231 "Mechanical vibration and shock." Technical committees other than TC 10, TC 122, TC 231, and TC 336 are omitted for clarity.

TABLE 6.2
Stages in the Production of a CEN Standard

Stage	Description	Deliverable	Abbreviation
1. Work definition	New work item proposed	New project proposal form	
2. Questionnaire procedure	Determination of actions required to distribute a reference document	Preliminary questionnaire (for new items) or updating questionnaire (for a revision)	
3. Technical body stage	Drafting of the European Standard	Working document (first draft of European Standard)	
4. CEN enquiry	To obtain and handle national comments	Final draft of the draft European Standard	prEN
5. Formal voting	Formal voting on the prEN	Final text of the European Standard	
6. Finalization and printing	Preparation of definitive language version of the EN and finalization of all translations	European Standard	EN
7. National implementation	Adoption as a national and international standard (depending on vote of appropriate committee)	National Standard International Standard	e.g., BS EN ISO

of regional (e.g., CEN) or international (ISO) standards. Therefore, the use of nationally standardized procedures is declining in favor of internationally standardized procedures.

Griffin (1990) lists almost 150 human-response-to-vibration standards published by national bodies, with a note that it is not a comprehensive list. Since then, it is likely that the number of standards has increased. Many national standards are cited in the literature; however, the most commonly cited national standards for human response to vibration are ANSI (American National Standards Institute), BSI (British Standards Institution), DIN (Deutsches Institut für Normung, Germany), and JISC (Japanese Industrial Standards Committee).

Many national standards are direct adoptions or translations of other documents. To avoid misinterpretation due to translation, the original (usually English) version is occasionally presented in parallel with a version in the native language (e.g., the Swedish Standard, SS ISO 8662-1, 1989).

Although this section will continue with discussion of only the BSI, equivalent national organizations throughout the world have also made important contributions and many use similar structures and methods.

6.2.3.1 British Standards Institution (BSI)

The British Standards Institution (BSI) was the first national standards body (established in 1901) and remains one of the most active. It is based in Chiswick, London. Committees work in the national interest and are comprised of individuals representing groups with an interest in the work. One important aspect of BSI's activity is the development of CEN and ISO standards and providing experts to attend international committees. Approximately 45% of newly introduced standards are adoptions of CEN standards, and 45% are adoptions of ISO standards, the remainder being purely national standards. At the British Standard level, a document may have a triple reference code indicating its adoption at international, European, and national levels (e.g., BS EN ISO 5349-1: 2001 Mechanical vibration — Measurement and evaluation of human exposure to hand-transmitted vibration — Part 1: General requirements).

BSI committees are categorized according to business sector. Most human vibration standards are developed within the "General Mechanical Engineering" standards committee under the remit of technical committee GME/21, "Mechanical vibration and shock." This committee is split into seven subcommittees, where GME/21/6 is human exposure to mechanical vibration and shock (Figure 6.4). Subcommittees can be subdivided into panels. For example, GME/21/6 has five panels; GME/21/6/5 is whole-body vibration. A further panel of relevance is GME/21/2/1: Instrumentation for the measurement of vibration and shock applied to human beings.

The independent development of British standards follows a similar process to those for international and European standards comprising new work items, draft standards, and a public consultation. However, the use of this procedure is becoming rare, with most BSI activity focusing on the development of standards for ISO or CEN.

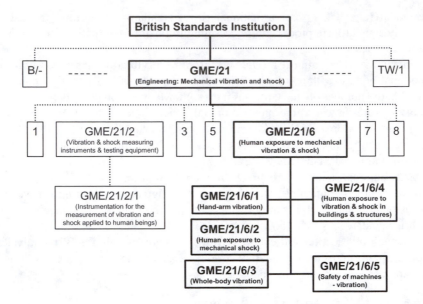

FIGURE 6.4 Organizational structure of BSI GME/21/6 Human exposure to mechanical vibration and shock. Technical committees other than B/-, GME/21 and TW/1 are omitted for clarity.

6.3 LEGAL POSITION OF STANDARDS

Standards are, in principle, documents designed to provide guidance and advice. Compliance is voluntary, and responsibility lies with the user to ensure that the standard is applied correctly and is the most appropriate for their application. Standards are most beneficial if all users apply them so that there is a common methodology or protocol.

Despite standards not being mandatory in themselves, it is possible for other mandated documents to refer to them. For example, legislation could state that a product must meet certain criteria when tested according to a standard. In this case, the standard is used as a tool by the legislator. It is not the standard that is legally binding, but the legislation that uses the standard. Other occasions when standards become binding include situations where the use of a standard is mandated according to a contract.

Within civil law, standards can also be used as evidence. If a standard exists specifying guidelines for hazard reduction (e.g., Published Document PD 6585-2: Hand-arm vibration — guidelines for vibration hazards reduction — part 2: management measures at the workplace, British Standards Institution, 1996), then employers might be considered negligent if they did not follow the guidance. Those assessing vibration exposures for a civil claim should also follow standardized methods for measurement as this is generally considered a good practice within the courts. If the investigator considers that alternative techniques not defined in the standard are superior, then these can also be reported in addition to the standardized method.

Standards are designed to provide a consensus of the best advice at their time of publication. The state of scientific knowledge is constantly changing, and as a result, standards are periodically updated if there is an improvement to be made or if the document requires correction. It is therefore possible that there is an apparent conflict either between standards or between a standard and what might be considered (at least by some) as a superior method. In this case, the investigators must use their judgment to select the appropriate methodology. Occasionally, contractual requirements specify that an outdated (i.e., superseded) standard must be used. Generally, this should be avoided unless there is a sound reason otherwise.

Even if the correct standard is used for an assessment, and it indicates a satisfactory situation, it is possible that a risk remains (either due to lack of power of the methods used or risks from other factors). Therefore, BSI adds a disclaimer to the standards: "Compliance with a British Standard does not of itself confer immunity from legal obligations" (from BS 6841, 1987).

The current version of a standard should be used for vibration assessment. Most standardization bodies have electronic databases that enable the status of any version of any standard to be checked. The original version of a standard should also be referred to rather than relying on a synopsis supplied by a vibration-measuring equipment supplier, seen in a scientific article, or even described in a book.

6.4 STANDARDS FOR WHOLE-BODY VIBRATION MEASUREMENT AND ASSESSMENT

The most well-known standards for measurement and evaluation of human exposure to whole-body vibration are BS 6841 (1987) and ISO 2631-1 (1997). Although BS 6841 (1987) was developed as a British standard, it has been used extensively (often in preference to ISO 2631-1, 1997) outside Britain. Conversely, ISO 2631-1 (1997) has seen a gradual increase in use within Britain since its publication, despite the concurrently accepted British standard. The methods described in the two standards will not give identical results, except in contrived circumstances. This example highlights the problem of the existence of more than one accepted method and the benefit of harmonization.

6.4.1 BS 6841: Guide to Measurement and Evaluation of Human Exposure to Whole-Body Mechanical Vibration and Repeated Shock (1987)

BS 6841 (1987) replaced the BSI Draft for Development DD32 (1974) and was designed to consolidate methods that were already being used in the industry but were only specified in a draft revision of ISO 2631 (Griffin, 1990). The scope of the standard includes methods for quantifying vibration and repeated shocks in relation to human health, interference with activities, discomfort, the probability of vibration perception, and the incidence of motion sickness. Much of the information often quoted from the standard is contained in the appendices, which are provided for information only and do not, technically, form part of the standard itself.

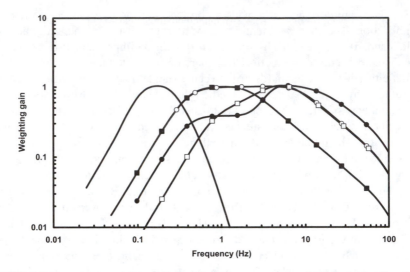

FIGURE 6.5 Frequency weightings for application to translational whole-body vibration as defined in BS 6841 (1987). Weightings shown are W_b (–●–), W_c (–○–), W_d (–■–), W_f (—), and W_g (–□–).

The standard states that vibration measurements on compliant seats should be made using an accelerometer mounted in a device such as an SAE pad (see Subsection 5.9.3) placed on the seat surface. The standard also gives guidance on how to assess vibration measured at the seat back and at the feet. Acceleration signals should normally be frequency weighted using W_b for seat vertical, W_c for backrest fore-and-aft, and W_d for horizontal vibration at the seat (Figure 6.5). For assessments of hand control and vision, the frequency weighting W_g should be used in preference to W_b. For motion sickness, W_f should be used.

To evaluate the effects of vibration on health, the root-mean-square (r.m.s., see Subsection 5.8.2) of the frequency-weighted acceleration signals should be calculated if the signals are of constant magnitude (i.e., stationary) and have a crest factor of less than 6. The estimated vibration dose value (eVDV) should then be calculated. If signals are not stationary, or if the crest factors are greater than 6, then the vibration dose value (VDV) should be calculated. Axes should be combined using a 4th power method. The standard defines a threshold of a VDV of 15 $m/s^{1.75}$ where severe discomfort occurs and states that: " ... *there is currently no consensus of opinion on the precise relation between vibration dose values and the risk of injury ... it is reasonable to assume that increased exposure to vibration will be accompanied by increased risk of injury.*" The standard is therefore careful to not present the 15-VDV threshold as a safe or unsafe border but a general indicator.

The process for calculating the VDV for application to health is relatively straightforward. If measurements are reported as being made according to BS 6841 (1987), then there is little scope for confusion. A simple flowchart can be followed to obtain the required result (Figure 6.6).

The mathematical formulae defining VDV allow for the equivalent r.m.s. acceleration magnitudes to be calculated for various VDVs. This is expressed graphically

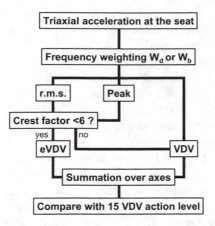

FIGURE 6.6 Method of evaluation and assessment according to health criteria in BS 6841 (1987). [Adapted from Griffin (1998a).]

in BS 6841 (1987) and allows for an illustration of the time dependency of VDV (Figure 6.7). For assessments according to VDV (or eVDV), a doubling of r.m.s. acceleration magnitude results in a reduction of equivalent exposure time by a factor of 16.

For evaluation of the effects of vibration on activities and performance, the r.m.s. magnitudes of vertical acceleration should be weighted using W_g. The relative effects

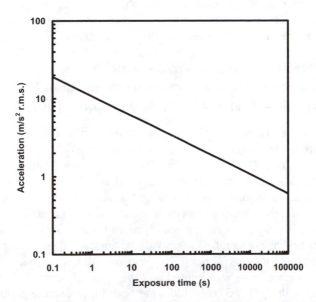

FIGURE 6.7 Root-mean-square acceleration (r.m.s.) corresponding to a vibration dose value (VDV) of 15 m/s$^{1.75}$ for exposure durations from 0.1 to 100,000 s. A doubling of r.m.s. acceleration magnitude results in a reduction of exposure time by a factor of 16 for the same VDV.

of different environments can be compared, but it is not possible to use BS 6841 (1987) to provide definitive limits on vibration for performance. However, for accurate hand positioning or resolution of precise visual details, the weighted acceleration magnitude in any axis should not exceed 0.5 m/s^2 r.m.s.

BS 6841 (1987) provides guidance on vibration discomfort and perception. Although assessments can be made just using triaxial–translational acceleration measurements on the seat surface, guidance is also provided to include rotational vibration and vibration at the feet and backrest. As for the assessments with regards to health, r.m.s. methods can be used for assessing low crest–factor exposures. If the crest factor exceeds 6, then VDV should be used. These are useful for determining the relative discomfort from different signals, but the standard states that other contextual factors dictate the acceptability of the vibration. According to BS 6841 (1987), 50% of alert and fit persons can detect a signal with a weighted acceleration magnitude of 0.015 m/s^2 peak.

The standard also provides guidance on the assessment of low-frequency vibration with respect to the incidence of motion sickness. Vertical vibration exposures should be weighted using W_f and assessed using the motion sickness dose value (MSDV$_z$, see also Subsection 3.4.1). For a mixed population, the percentage of unadapted persons who are likely to vomit can be estimated using:

$$\text{Vomiting incidence } \% = \text{MSDV}_z$$

BS 6841 (1987) provides guidance that is generally clear with little scope for confusion. Users need to take care to ensure that the correct frequency weightings are used (e.g., do not confuse W_b with W_g for seat surface measurements), and that crest factors are monitored to ensure that the correct method of assessment is used.

6.4.2   ISO 2631: MECHANICAL VIBRATION AND SHOCK — EVALUATION OF HUMAN EXPOSURE TO WHOLE-BODY VIBRATION — PART 1: GENERAL REQUIREMENTS (1997)

ISO 2631 (1997) replaced an earlier version of the standard (ISO 2631, 1985) that contained different frequency weightings and different criteria (for a comparison of the 1985 version of ISO 2631 with the 1997 version, see Griffin, 1998a). It also replaced ISO 2631-3 (1985) which was then withdrawn. ISO 2631 currently contains three parts including guidance on evaluating vibration exposure in buildings (ISO 2631-2, 1989) and the effects of vibration and rotational motion on passenger and crew comfort in fixed-guideway transport systems (ISO 2631-4, 2001). At the time of writing, Part 5 has been drafted to provide additional guidance on assessment of vibration containing multiple shocks. This section will only consider Part 1 of the standard as this is the most widely used and the most relevant.

The scope of the ISO 2631-1 includes methods for measuring periodic, random, and transient whole-body vibration, with annexes that provide guidance on the interpretation of the measurements. The first annex provides a definition of the frequency weightings and is *normative* (i.e., it forms part of the standard). The other four annexes are *informative* and therefore do not form part of the standard itself.

Nevertheless, the content of these annexes forms the focus of much discussion regarding the standard.

The standard states that vibration should be measured at the interface between the human body and the vibration source. Measurements on a seat should be made beneath the ischial tuberosities (the bony points that can be felt if one sits on one's hands) using an accelerometer mount (e.g., an SAE pad; see Subsection 5.9.3), provided the seat is compliant. The standard also gives guidance on how to approach vibration measurements made at the seat back and at the feet. Acceleration signals should normally be frequency weighted using W_k for seat vertical, W_c for backrest fore-and-aft, and W_d for horizontal vibration at the seat. For motion sickness, W_f should be used.

Evaluation of the effects of vibration on health, according to ISO 2631 (1997), is determined using the frequency-weighted r.m.s. for each axis of translational motion on the supporting surface if the crest factor is less than 9. Assessments are made independently in each direction, and horizontal vibration is multiplied by a scaling factor of 1.4. The overall assessment is usually carried out according to the worst axis of frequency-weighted r.m.s. acceleration (including multiplying factors). If two axes have comparable magnitudes, then these can be combined using the vector sum method, although this is not mandated and guidance is not given as to how to interpret "comparable." If crest factors exceed 9, then two alternative methods of assessment are suggested: the maximum transient vibration value (MTVV) and the VDV. The MTVV is defined as the highest value of the running r.m.s. for the measurement period, where the running r.m.s. is a measure of the acceleration magnitude in the previous second. Even if VDV or MTVV are used, r.m.s. should be reported.

Two "health guidance caution zones" are included in ISO 2631 (1997) to assist with interpreting the worst axis of the frequency-weighted r.m.s. acceleration (Figure 6.8). Two zones are provided as they are derived from r.m.s. and VDV approaches. The standard states that: *"For exposures below the zone, health effects have not been clearly documented and/or objectively observed; in the zone, caution with respect to potential health risks is indicated and above the zone health risks are likely."* The zones coincide for durations of about 4 to 8 h, and the standard warns against using the zones for shorter durations. Indeed, for exposure durations between about 5 and 30 min, it is possible to exceed the limits of the zone according to one method and not reach the zone for the other! A doubling of r.m.s. acceleration magnitude results in a reduction of exposure time by a factor of 16 for one of the methods and a reduction of exposure time by a factor of 4 for the other. For assessments according to VDV, the health guidance caution zone has upper and lower bounds at 8.5 and 17 $m/s^{1.75}$, respectively. There is no equivalent zone for MTVV.

The process of making a health assessment according to ISO 2631 (1997) is complex and can be confusing. If two individuals were asked to make measurements according to ISO 2631 (1997), a wide range of results could be attained. The range of alternative methods of assessment that are allowable is not helpful. However, at the core of the standard is the method of using frequency-weighted r.m.s., and this is the primary method that most users apply (VDV is sometimes used for signals with high crest factors). Many of the superfluous options that are allowable could

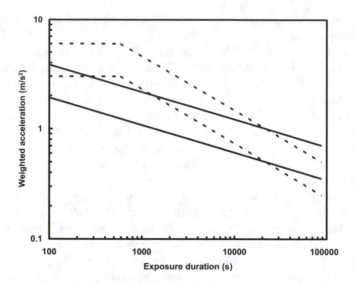

FIGURE 6.8 The two "health guidance caution zones" given in ISO 2631 (1997).

be deleted in future versions without impacting the majority of users. A simplified flowchart that can be followed for most assessments is shown in Figure 6.9.

For evaluation of the effects of vibration on comfort and perception according to ISO 2631, the frequency-weighted r.m.s. acceleration is determined for the three translational axes on the seat. Also, methods to include rotational vibration and vibration at the feet and seat back are described. The same frequency weightings are used for health assessments (although the standard also refers to using W_b for assessments of rail vehicles). If vibration measurements cannot be made at the backrest, then multiplying factors of 1.4 for the horizontal axes and 1 for the vertical axis should be used for comfort. However, for perception, multiplying factors of 1

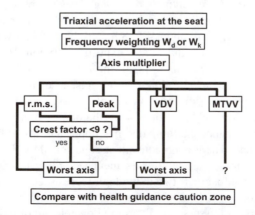

FIGURE 6.9 Simplified method of evaluation and assessment of whole-body vibration according to health criteria defined in ISO 2631 (1997).

TABLE 6.3
Approximate Indications of Likely Reactions to Various Magnitudes of Overall Vibration Total Values in Public Transport as Stated in ISO 2631-1 (1997)

Frequency-Weighted Vibration Magnitude (Root-Sum-of-Squares)	Likely Reaction in Public Transport
Less than 0.315 m/s^2	Not uncomfortable
0.315 m/s^2 to 0.63 m/s^2	A little uncomfortable
0.5 m/s^2 to 1 m/s^2	Fairly uncomfortable
0.8 m/s^2 to 1.6 m/s^2	Uncomfortable
1.25 m/s^2 to 2.5 m/s^2	Very uncomfortable
Greater than 2 m/s^2	Extremely uncomfortable

Note: The magnitudes and likely reactions are identical to those found in BS 6841 (1987), although BS 6841 does not specify any application area.

---

are used on all translational vibration axes. If the crest factor exceeds 9, then VDV or MTVV should be used. Vibration from all axes should be combined. The standard states that environmental and situational factors greatly influence the acceptability of any vibration magnitude, although criteria are included in the text (Table 6.3). According to ISO 2631 (1997), 50% of alert and fit persons can detect a W_k weighted acceleration with a magnitude of 0.015 m/s^2 peak.

Guidance with respect to predicting the incidence of motion sickness is also included in the standard. The methods are nominally identical to those found in BS 6841 (1987).

International Standard ISO 2631 (1997) generally provides guidance that is, at best, nonspecific and, at worst, confusing. There are many clauses that leave decisions open to the judgment of the investigator, and can result in misleading conclusions (Lewis and Griffin, 1998). Other clauses describe how to calculate a value (e.g., MTVV), but then no guidance is given on how to interpret it. One can envisage a simplified, more user friendly revision of the standard that leaves less to the judgment (or misjudgment) of the investigator and yet is entirely coherent with the spirit (if not the letter) of the methods specified in the 1997 version.

A user of the ISO 2631 (1997) is advised to clearly record and report the methods used, as stating "in accordance with ISO 2631" is often unspecific. Since the current version's publication in 1997, those using ISO 2631 have tended to keep to measurements of r.m.s. with occasional additional reports of VDV. The axis multipliers are an important aspect of the standard that must be correctly used depending on the application.

6.4.3 COMPARISON OF BS 6841 (1987) AND ISO 2631 (1997)

In many respects, British Standard BS 6841 (1987) and International Standard ISO 2631 (1997) are similar. For motion sickness assessments the standards are substan-

TABLE 6.4

Key Differences between British Standard BS 6841 (1987) and International Standard ISO 2631-1 (1997)

BS 6841	ISO 2631-1
Uses W_b frequency weighting for vertical vibration	Uses W_k frequency weighting for vertical vibration
No multiplication factors required for measurement of seat surface vibration	Horizontal vibration is scaled by a factor of 1.4 for measurements of seat surface vibration
Use of 15 VDV as criteria for health risk	Use of two "health guidance caution zones" as criteria for health risk
Use of combined axes for assessments	Use of "worst axis" for assessments, with option to use combined axes
Crest factor of 6 used as threshold for unreliability of r.m.s. methods	Crest factor of 9 used as threshold for unreliability of r.m.s. methods

tially identical; for perception and comfort evaluations they are broadly similar. The methods of acquiring data and locations for measurement are alike. It is also possible to interpret ISO 2631 so that most vibration analyses can be carried out with a similar approach to BS 6841.

However, notwithstanding the differences in clarity of the two standards, there are five core areas where BS 6841 and ISO 2631 differ. These are the different vertical frequency weightings, the multiplication factors for horizontal vibration in ISO 2631, the use of VDV or r.m.s. as the criteria for assessment of health, whether assessment should be made using combined axes or the worst axis only, and the value for the crest factor at which r.m.s. becomes unreliable. These are summarized in Table 6.4.

Subsection 2.2.4 describes the differences between the frequency weightings W_b and W_k. Although the weightings are broadly similar in shape, if vibration is dominated at those frequencies where there is a difference (e.g., 25% difference at frequencies below 3 Hz), then, of course, the results will be scaled accordingly. Therefore, wherever possible, comparisons between measurements made using different frequency weightings should be avoided. Lewis and Griffin (1998) compared measurements made with the two weightings for nine different vehicles and showed a maximum difference of 17% between the frequency-weighted r.m.s.s, with measurements made using W_k exceeding those made using W_b for all motorized vehicles. Rarely will the use of different frequency weightings alter the rank orders of analyzed data if each axis is considered separately.

The differences between the results obtained using alternative standards due to the multiplication factor for horizontal vibration depends on the relative contribution of the horizontal component to the total vibration exposure. If the vibration is solely in a horizontal axis (or the worst axis is horizontal), then ISO 2631's axes multiplier will increase the result by 40%. However, if using ISO 2631 and the worst axis is vertical, then the multipliers have no effect as the data from the horizontal axes are ignored in the final assessment. It is unfortunate that if either r.m.s. or VDV data are used to estimate health risk based on either BS 6841 or ISO 2631, the results will be different for each of the orthogonal axes. For vertical vibration the frequency weightings are different; for horizontal vibration axes, multipliers apply for ISO 2631 but not for BS 6841.

Both ISO 2631 (1997) and BS 6841 (1987) use r.m.s. as the basis for measurement of signals with crest factors below 6. In these cases the conversion from r.m.s. to eVDV in BS 6841 alters the time dependency from one where a doubling of magnitude results in a 4-fold decrease in equivalent exposure to one with a 16-fold decrease in equivalent exposure. Retaining r.m.s. throughout the analysis process is one of the reasons why one of the health guidance caution zones in ISO 2631 retain the four-fold relationship between doubling of vibration magnitude and equivalent exposure time. VDV is better suited to the analysis of signals with repeated or occasional shocks than r.m.s. (i.e., signals with a high crest factor), as the influence of these events do not decay as subsequent time elapses (see Subsection 5.8.2).

So, should an investigator use BS 6841 (1987) or ISO 2631 (1997)? Perhaps the safest approach would be to use a data acquisition system and to analyze acquired acceleration according to both standards. Then, if action values or caution zones are exceeded for both data sets, a clear conclusion can be drawn. Often, this approach is impractical or impossible, in which case regional or industrial sector standard practice should be followed. With the phasing in of the EU Physical Agents (Vibration) Directive across Europe (which specifies the use of ISO 2631), it is likely that the use of BS 6841 (and therefore W_b) will decline in favor of ISO 2631 (and therefore W_k). The complexity, confusing approach, and content of ISO 2631 will not improve with an increased user population. Indeed, considering that the majority of this extended user group will be new to the field, scope for increasing the confusion is substantial.

Whichever standard is selected by an investigator, the results should be presented so that there is transparency regarding the methods used. Ideally, at least the frequency-weighted r.m.s., eVDV, VDV, and crest factor should be reported for each axis (with and without the inclusion of the multiplying factors), in addition to summary and overall results. Thus the reader should be able to calculate results using an alternative method, an alternative interpretation of the same standard, or even estimate results if a different standard was used.

For a more in-depth comparison of ISO 2631 and BS 6841, see Griffin (1998a) and Lewis and Griffin (1998).

6.5 STANDARDS FOR DETERMINING VIBRATION EXPOSURES IN SPECIFIC VEHICLES

6.5.1 ISO 5008: AGRICULTURAL WHEELED TRACTORS AND FIELD MACHINERY — MEASUREMENT OF WHOLE-BODY VIBRATION OF THE OPERATOR (2002)

ISO 5008 (2002) has also been adopted as a national standard (e.g., BS 5008, 2002). It replaces an earlier version of the standard that was published in 1979. Unlike most other human vibration standards, it was developed under the auspices of ISO

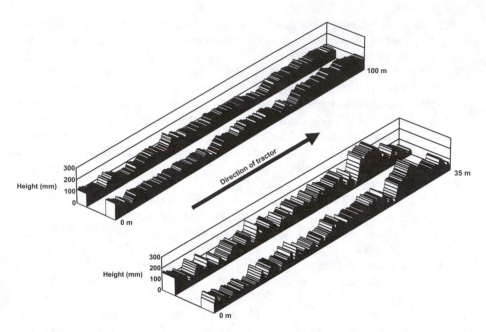

FIGURE 6.10 Elevation and profile of 35-m ("rougher") and 100-m ("smoother") tracks for testing whole-body vibration exposures in tractors, as defined in ISO 5008 (2002). (*Note:* Horizontal axes are not to scale.)

TC 23 (tractors and machinery for agriculture and forestry) rather than ISO TC 108. The purpose of the standard is to specify methods that should be used to assess the whole-body vibration exposure in tractors. The general methods of vibration measurement specified in the standard are in accordance with ISO 2631 (1997).

The distinctive aspect of ISO 5008 (2002) is that it defines two artificial test tracks designed to provide a standardized surface that will produce vibration at the driver's seat. The tracks must be constructed of a series of concrete, steel, or wooden blocks of heights specified in the standard. Therefore, nominally identical tracks can be built in different locations, and similar results should be obtained when testing a tractor at any test site. One track is a 35-m "rougher track" defined at 8-cm intervals, and the other is a 100-m "smoother track" defined at 16-cm intervals (Figure 6.10). The tracks are defined such that different profiles exist for the left and right wheels. Test tractors are driven along the tracks at 10, 12, and 14 km/h and at 4, 5, and 7 km/h for the smoother and rougher tracks, respectively, with triaxial measurements of vibration made at the driver's seat. Five repeat measures are required at each speed for each operator tested (either one operator of 75 kg ± 5 kg for a standard type of tractor with a laboratory-tested seat in accordance with ISO 5007, or two operators, one of 52 to 55 kg and one of 98 to 103 kg, if the tractor is not a standard type or the seat has not been tested in the laboratory). The standard also provides guidance on testing tractors in the field.

No pass or fail criteria are specified within ISO 5008 (2002); its purpose is to provide a standardized test, rather than guidance on interpretation of the results.

One problem with this standard is that, although the speeds specified are appropriate for traditional tractors, some new designs include suspension at the axles. This provides a better isolation for the machine and driver and enables the tractor to be driven at higher speeds across rough terrain. It is possible to drive such machines across these test tracks at speeds of 30 km/h (Scarlett et al., 2002); for such tractors the slow-speed tests might not be representative of real use.

6.5.2 ISO 10056: MECHANICAL VIBRATION — MEASUREMENT AND ANALYSIS OF WHOLE-BODY VIBRATION TO WHICH PASSENGERS AND CREW ARE EXPOSED IN RAILWAY VEHICLES (2001)

ISO 10056 (2001) has also been adopted as a national standard (e.g., BS 10056, 2001). It is designed to supplement ISO 2631 (1997) such that it can be applied to the railway environment, especially as railway vibration has some specific characteristics. It provides guidance on measurement and analysis of vibration on board railway vehicles but does not include methods for assessing the effects of the vibration or specific criteria. The standard is not designed for application to railway-induced motion sickness.

ISO 10056 (2001) states that measurements should be made on the floor and possibly at the seat. As railway-vehicle vibration varies with location, measurements should be made on the floor over the bogie centers, at the center of the vehicle body, and on the vestibule floor (i.e., the enclosed area at the end of a passenger car). Seat vibration should be measured at multiple locations. The standard requires measurements to be at least 20 min in duration and analyzed in 5-min segments. Analysis should be carried out in accordance with ISO 2631 (1997) to generate frequency-weighted r.m.s. accelerations. In addition, the weighted r.m.s. is calculated every 5 s. These data are then plotted as a histogram, indicating the percentage of time that any vibration magnitude was present. Spectral analysis should also be carried out. The standard gives example data showing power-spectral densities of vibration on each axis.

One of the unusual features of railway motion is that, unlike most other forms of land transport, nominal stationary signals exist for long periods of time. The histograms described in ISO 10056 (2001) are useful indicators of the variability of the vibration. This principle can be applied for assessing the variability of vibration exposures in other environments where measurement is possible for long periods of time (e.g., aircraft, road haulage).

6.6 STANDARDS FOR DETERMINATION OF WHOLE-BODY VIBRATION EMISSION VALUES FOR MOBILE MACHINERY

Employers have the responsibility to ensure that the health and safety of their workforce is not compromised. Therefore, steps should be taken to minimize the risks associated with any operation. Risks arising from whole-body vibration can be reduced by careful selection of machines during purchase, such that a low-vibration-emission machine is acquired. Combining these data with the duration of expected use allows a prediction of the exposure for any individual to be made.

Most machines can be used in a variety of situations that largely depend on the requirements of the purchaser, resulting in a broad range of possible vibration magnitudes. Therefore, it is impossible for a manufacturer to state the vibration emission for every particular application, and it is difficult to give an accurate estimate for any application. Nevertheless, purchasers demand such information, and therefore manufacturers face a dilemma in reporting vibration magnitudes in machines: Do they report optimal conditions, thereby enhancing their commercial position, or do they report average (or even the worst) conditions and run the risk that a competitor will use a different strategy and obtain an unfair commercial advantage? It is beneficial for purchasers to have the confidence that manufacturers have used similar test methods to obtain their vibration emission values; hence, there is a need for a standardized test procedure.

6.6.1 EUROPEAN STANDARD EN 1032: MECHANICAL VIBRATION — TESTING OF MOBILE MACHINERY IN ORDER TO DETERMINE THE VIBRATION EMISSION VALUE (2003)

EN 1032 (2003) has also been adopted as a national standard across Europe (e.g., BS 1032, 2003). The standard superseded an earlier version that was published in 1996. It is designed to support the EU Machinery Safety Directive (Council of the European Communities, 1998) which states that the instruction handbook of a machine must contain the weighted r.m.s. acceleration to which the body is subjected if it exceeds 0.5 m/s², tested according to the most appropriate method (see also Section 8.4). EN 1032 provides a basis for the development of other standards to define the most appropriate method for type testing, such that the results are representative of those obtained in real working conditions. The earlier version of the standard only considered whole-body vibration exposures in machines whereas the 2003 version also includes a consideration of the hand-transmitted vibration that might affect the operator.

EN 1032 (2003) does not include limits or recommended methods for interpretation of results from emission tests, but allows for comparison of results obtained from different machines. It specifically excludes the evaluation of rotational motion and backrest vibration.

The approach of EN 1032 (2003) is broadly compatible with ISO 2631 (1997) such that it recommends the use of a semirigid accelerometer mount for measurements on compliant seats (i.e., an SAE pad). Also, measurements are to be reported as m/s² r.m.s., frequency weightings W_k and W_d should be used, and the 1.4 axis multipliers for horizontal vibration should be applied. If vibration is dominated by one axis, it should be reported; if there is no dominant axis then axes can be combined using the root-sum-of-squares method. One important inclusion in EN 1032, which is an improvement on ISO 2631, is that a criteria is provided to establish whether any axis is dominant. The standard states that an axis is dominant if the vibration from each of the other two axes is less than 66% of that in the dominant axis (including the axis multiplier for horizontal motion). Root-mean-square measures should be reported for sample durations that are at least as long as the duration for one complete work cycle. In addition, all measurements should last longer than 180 s for whole-body vibration emission tests.

The standard also specifies a method for the measurement of hand-transmitted vibration in mobile machinery. This might be transmitted to the operator from controls, levers, or steering wheels. The methods are broadly compatible with ISO 5349-1 (2001). Vibration measurements should be made as close to the normal position of the hand as possible. Triaxial acceleration should be measured and weighted using the W_h frequency weighting. The emission value is calculated as the sum of all axes of vibration (summed using r.s.s. methods). Measurements should last at least 12 s, and care should be taken to remove operator-induced elements in the data, such as steering wheel movements. To reduce the likelihood of such elements, measurements would normally take place on a straight course.

Vibration test codes developed in accordance with EN 1032 (2003) should ensure that measurements are accurate and reproducible with precisely defined operating conditions. A measure of the variability of the measurements should be included in the analysis. Measurements should include one condition that represents the 75th percentile of the vibration values obtained during the mode of operation that produces the highest vibration values in typical intended use of the machinery. The test code might also include the definition of an artificial test track or simplified test conditions if these are representative of the vibration emissions experienced in the normal working environment. Test codes can also mandate laboratory testing of parts of the machine (e.g., the driver's seat) if required.

Some types of machines (e.g., excavators) can be used with a wide array of attachments and tools. Where these would influence the vibration emission value, tests with a range of attachments might be required. This could prove to be a problem from a practical point of view, as manufacturers of attachments are not necessarily the manufacturers of the machines. Therefore, there is a question as to who should be responsible for testing the vibration emission of the machine if it has been fitted with a nonstandard attachment.

Despite the previous version of EN 1032 (1996) being implemented for some years, there were very few CEN standards that have been developed from it. One problem is that most types of machines sold in Europe can be used, to some extent, in a number of situations. For example, identical models of tracked excavators are used for quarrying, construction, and demolition, making it problematic to specify typical operation or one representative of the highest vibration values. There are also problems for machines that are generally used in one industry. For example, the vibration emission of a dump truck operating in a quarry will change, depending on weather conditions (e.g., frozen, hard, or soft ground), the type of material being extracted (e.g., hard rock, gravel, or clay), and the condition of the roadway (e.g., potholed, graded, or tarmac). ISO 5008 (2002) for testing of tractors was developed from the earlier version of the standard (1979) and has not been implemented as an EN standard, despite many aspects of it complying with EN 1032.

The lack of specific standards for individual vehicles means that manufacturers have the difficult task of complying with the Machinery Directive and must create their own methods for determination of the emission value. If these approaches differ

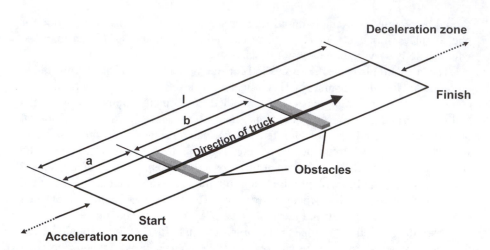

FIGURE 6.11 Schematic of test track for determining vibration emission values of industrial trucks as specified in EN 13059 (2002). Vehicle speed, obstacle height, and distances *a*, *b*, and *l* are defined for each type of truck.

between companies, then artificial systematic differences could be indicated between manufacturers which could mislead potential purchasers.

6.6.2 EN 13059: SAFETY OF INDUSTRIAL TRUCKS — TEST METHODS FOR MEASURING VIBRATION (2002)

EN 13059 (2002) has also been adopted as a national standard across Europe (e.g., BS 13059, 2002). The standard is in general accordance with EN 1032. It is designed to support the Machinery Directive 98/37/EC such that the methods specified in the standard can be used as declared values, and that similar results would be obtained irrespective of where the tests are carried out. It cannot be used to estimate operator vibration-exposure magnitudes in normal work conditions.

The industrial trucks to be tested are loaded and driven across a test track (Figure 6.11). The track must comprise a relatively smooth surface with two 150 mm-wide obstacles of a height and separation dependent on the type of truck (Table 6.5). An interesting inclusion in this standard is guidance for developing new measurement methods, should new categories of industrial trucks be launched.

The vibration emission value is reported as the frequency-weighted r.m.s. acceleration measured on the seat or floor, depending on whether the truck is used with seated or standing operators (or both). Values must be rounded to the nearest 0.1 m/s^2 r.m.s.

6.7 STANDARDS FOR TESTING VEHICLE SEATS IN THE LABORATORY

Operator seats can be considered a form of personal protective equipment (PPE) for many machines. A seat is also one of the least expensive parts of the machine to

TABLE 6.5
Truck Categories and Associated Test Conditions for Determining Emission Values of Industrial Trucks according to EN 13059 (2002)

Category	1	2	3	4	5
Truck Type	Platform Trucks, Trucks Rider Controlled, etc.	Reach Trucks, Articulated Trucks, etc.	Straddle Trucks, Counterbalanced Trucks, etc.		
Wheel diameter (mm)	≤ 200	> 200	≤ 645	> 645 and ≤ 1200	> 1200 and ≤ 2000
Length of test track, l (m)	15	25	25	25	25
Height of obstacles (mm)	5	5	8	10	15
Distance from start to obstacle 1, a (m)	4	5	5	5	5
Distance between obstacles, b (m)	6	10	10	10	10
Speed (km/h)	5	7	10	10	10

replace, and so there is a market for replacement seats for machines where vibration is considered to be a problem. As poor suspension seats can amplify rather than attenuate the vibration, it is appropriate that there are seat test codes to ensure a minimum performance (and therefore quality) of seats.

Current vibration test codes for seats are based on criteria using SEAT values and tests of the damping performance of the seat under worst-case conditions. They are conducted in the laboratory. At the time of writing, other test codes are being drafted that include tests of the dynamic performance of the end-stop buffers. In addition to vibration tests, further criteria are specified in other standards, such as seat strength, dimensions, and durability. These are beyond the scope of this chapter.

6.7.1 ISO 10326-1: Mechanical Vibration — Laboratory Method for Evaluating Vehicle Seat Vibration — Part 1: Basic Requirements (1992)

ISO 10326-1 (1992) has also been adopted as a European standard (EN 30326-1, 1994) and as a national standard (e.g., BS 30326-1, 1994). It is designed to provide a basis for the development of test codes for the laboratory evaluation of the dynamic characteristics of seats used in vehicles and mobile off-road machinery. ISO 10326-1 does not define the details of these test codes.

ISO 10326-1 describes two types of laboratory tests that can be specified for vehicle seats. One test is designed to evaluate the dynamic performance of the seat while exposed to simulated typical vehicle vibration. The second test evaluates the performance of the seat when exposed to transients or severe shocks. The vibration is measured at the base of the seat (i.e., on the moving platform of the simulator), the surface of the seat, and also the backrest (if applicable). Seat measurements are made with a standard accelerometer mounting pad. Before any testing takes place, suspension seats must be run-in by loading the seat with a 75-kg rigid mass and exposing the seat to sinusoidal vibration at the seat's resonant frequency such that 75% of the seat travel is used. The run-in period should be long enough for the performance of the seat to stabilize. For the tests themselves, two subjects are required: one with a mass equal to the 5th percentile and the other with a mass equal to the 95th percentile of the intended user population.

The simulated vehicle test (seat transmissibility test) exposes the seat to motion in the laboratory that is representative of that which would be experienced in severe, but normal, use. The characteristics of the test signal are defined, as is the required r.m.s. vibration magnitude for the tests. An additional test using a swept sine signal (a sine wave that steadily increases in frequency) can also be specified in some test codes to enable the seat transmissibility to be determined. Three repeats are required for each subject where the frequency-weighted vibration on the surface of the seat is measured. These data are used to determine the SEAT value (see Subsection 2.5.2). The seat passes the test if the SEAT value is less than an acceptable value specified in the standard that is based on ISO 10326-1.

ISO 10326-1 specifies a damping test whereby suspension seats are excited at resonance. For this test, the seat is occupied by an inert mass of 75 kg. The test is defined to enable the transmissibility of the seat at a resonance to be established. The test is repeated three times.

ISO 10326-1 (1992) has been used to develop a range of seat test codes including those for industrial trucks and earthmoving machinery.

6.7.2 ISO 10326-2: MECHANICAL VIBRATION — LABORATORY METHOD FOR EVALUATING VEHICLE SEAT VIBRATION — PART 2: APPLICATION TO RAILWAY VEHICLES (2001)

ISO 10326-2 (2001) has also been adopted as a national standard (e.g., BS 10326-2, 2001). It is designed to enable tests to be completed to investigate the dynamic performance of railway seats that are provided for passengers and crew. Although some aspects are similar to the procedures defined in ISO 10326-1 (1992) there are additional tests and many differences.

ISO 10326-2 describes two tests that must be completed. The first test is used to characterize the seat transmissibility; the second test investigates the linearity of the seat's dynamic response with vibration magnitude. For each test, nine acceleration measurements are made: a triaxial measurement on the surface of the simulator platform, a triaxial measurement on the surface of the seat, and a triaxial measurement at the backrest. In contrast to other human vibration standards, ISO 10326-2 specifies that accelerometers (used to measure the vibration on the seat surface) are mounted in a perforated, semirigid, and contoured "seat pan" that is shaped to approximate the contour of the seat–person interface on an occupied train seat. This device does not conform to the design of the SAE pad. Testing is performed using two test subjects with masses of about 55 kg and 90 kg, respectively. These tests are carried out separately, irrespective of the number of occupants who can use the seat simultaneously. Tests are carried out using vertical, lateral, and fore-and-aft vibration, sequentially.

The seat transmissibility test uses a broadband vibration signal (with components from 0.5 to 50 Hz) lasting at least 90 s. The magnitude of the signal must be 1.6 m/s^2 r.m.s. (unweighted). For the measurements of vibration at the seat and backrest, the frequency response and coherence functions are calculated (see Subsection 5.8.3). Two other values are calculated: the first (termed "transmissibility" in the standard) is the ratio of the unweighted r.m.s. vibration magnitudes measured on the seat and the shaker platform, the second value (termed "weighted transmissibility" in the standard) is similar to the SEAT value, but is calculated in each direction.

The sinusoidal excitation test specified in ISO 10326-2 (2001) uses the results of the frequency response functions to select those frequencies that show a resonant response. At these frequencies the seat is excited at 0.5 m/s^2 and at 1 m/s^2 using sinusoidal vibration. If there is a difference between the frequency response functions at these frequencies of more than 30% of the higher value, then nonlinear behavior is signified.

ISO 10326-2 (2001) provides no guidance on interpretation of any of the results that are obtained using the method. It is simply a technique for testing the dynamic performance of railway seats in the laboratory. It is difficult to see why a new device for mounting accelerometers needs to be defined, especially as ISO 10056 (see Subsection 6.5.2) advocates the use of the standard accelerometer mounting pad.

6.7.3 ISO 7096: EARTHMOVING MACHINERY — LABORATORY EVALUATION OF OPERATOR-SEAT VIBRATION (2000)

ISO 7096 (2000) has also been adopted as a European standard (with the addition of an extra annex; EN 7096, 2000) and as a national standard (e.g., British Standard BS 7096, 2000). It can also be referred to as BS 6912-17 (2000). It replaced the second edition of the standard ISO 7096 (1994). The standard is based on ISO 10326-1 but with specific application to seats intended for earthmoving machinery. It restricts its application to frequencies between 1 Hz and 20 Hz, and it specifies criteria by which a seat can be classed as having passed or failed the test.

There are nine categories of earthmoving machinery that are included in ISO 7096 (Table 6.6). These include tracked and wheeled vehicles, primarily comprising of wheel loaders, dozers, and dump trucks. Vibration spectra for the transmissibility tests are defined for each of the nine categories, and are termed EM1 to EM9 (Figure 6.12). The spectra are designed to simulate the vibration experienced in each of the machines. For example, vibration for category EM7 (compact dump trucks) is high magnitude and narrow band, centered on 3.2 Hz, whereas vibration for category EM6 (crawler trucks) is of a lower magnitude, broader band, and higher frequency. As it is easier to isolate higher frequency vibration, transmissibility acceptance criteria (Table 6.6) are less stringent for those categories that are dominated at low frequencies (e.g., EM1: articulated or rigid frame dump trucks over 4500 kg).

The damping test for earthmoving machinery seats are carried out at the seat resonance frequency and with a magnitude such that the peak-to-peak suspension displacement is approximately 40% of the total travel. The damping performance acceptance criteria are more stringent for seat categories EM1, EM2, EM3, EM4, and EM6 than for the other seats.

6.7.4 EN 13490: MECHANICAL VIBRATION — INDUSTRIAL TRUCKS — LABORATORY EVALUATION AND SPECIFICATION OF OPERATOR-SEAT VIBRATION (2001)

EN 13490 (2001) has also been adopted as a national standard across Europe (e.g., BS 13490, 2002). It is based on ISO 10326-1 (1992) but adapted specifically for seats intended for industrial trucks. Industrial trucks include forklift type trucks, as well as other vehicles for moving and stacking materials. The standard provides two criteria for assessing the seats: the SEAT value and the damping test, as defined in ISO 10326-1.

Similar to ISO 7096 (2000), the truck types are categorized and assigned one of a variety of test spectra with specific pass or fail criteria (Table 6.7). The four test spectra are labeled IT1 to IT4; classes IT2 (medium-sized straddle trucks) and IT3 (large trucks) have identical vibration spectra to those used to test compact

TABLE 6.6
Summary of Vehicle Categories and Acceptance Criteria for Evaluation of Seats Mounted in Earthmoving
Machinery, as Specified in ISO 7096 (2000)

Category of Vehicle	Description of Vehicle	Unweighted Vibration Magnitude for Transmissibility Test (m/s²)	Acceptable SEAT Value	Acceptable Damping Performance (Transmissibility at Resonance)
EM1	Articulated or rigid frame dumper > 4,500 kg	1.71	< 110%	< 1.5
EM2	Scraper without axle or frame suspension	2.05	< 90%	< 1.5
EM3	Wheel loader > 4,500 kg	1.73	< 100%	< 1.5
EM4	Grader	0.96	< 110%	< 1.5
EM5	Wheel dozer, soil compactor, backhoe loader	1.94	< 70%	< 2.0
EM6	Crawler dumper, crawler dozer, crawler loader ≤ 50,000 kg	1.65	< 70%	< 1.5
EM7	Compact dumper ≤ 4,500 kg	2.26	< 60%	< 2.0
EM8	Compact loader ≤ 4,500 kg	1.05	< 80%	< 2.0
EM9	Skid steer loader ≤ 4,500 kg	1.63	< 90%	< 2.0

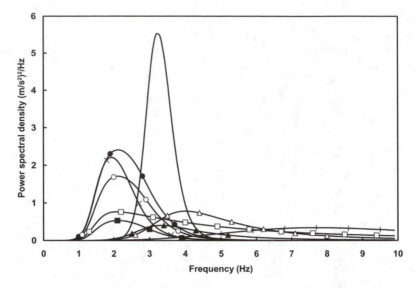

FIGURE 6.12 Power spectral densities of input vibration for transmissibility testing of nine categories of earthmoving machinery seats as specified in ISO 7096 (2000). EM1 (–×–), EM2 (–●–), EM3 (–○–), EM4 (–■–), EM5 (–□–), EM6 (+), EM7 (——), EM8 (–▲–), and EM9 (–△–).

loader trucks (EM8) and grader trucks (EM4) in ISO 7096. Damping tests are only required for large trucks and all-terrain industrial trucks.

6.8 CHAPTER SUMMARY

Whole-body vibration standards are developed by national, regional, and global committees representing bodies such as the International Organization for Standardization, European Committee for Standardization, and national bodies such as the British Standards Institution or the Deutsches Institut für Normung (Germany). Modern standards are usually endorsed by more than one institution such that international cooperation is encouraged. Standards are not legally binding in themselves, but legislation can require that a standard be adhered to. Alternatively, standards can be considered as best practice and in a claim for damages following vibration injury, noncompliance with a standard could be considered negligent.

BS 6841 (1987) and ISO 2631 (1997) are the most commonly used standards for assessing whole-body vibration. The two standards can be applied in similar ways, although there are important differences (e.g., frequency weightings, axis multipliers, consideration of single or multiaxis vibration). ISO 2631 leaves more options open to the investigator so that a measurement according to the standard can be executed and interpreted in a variety of ways that can change the ultimate conclusion of the tests. The popularity of BS 6841 is declining in favor of ISO 2631, as the international standard is referred to in other documents (e.g., the EU Physical Agents (Vibration) Directive).

TABLE 6.7
Summary of Vehicle Categories, Test, and Acceptance Criteria for Evaluation of Seats Mounted in Industrial Trucks, as Specified in EN 13490 (2001)

Category of Vehicle	Description of Vehicle	Unweighted Vibration Magnitude for Transmissibility Test (m/s²)	Peak Frequency for Transmissibility Test (Hz)	Acceptable SEAT Value	Acceptable Damping Performance (Transmissibility at Resonance)
IT1	Trucks with wheel diameter < 645 mm	1.58	5.0	< 70%	exempt
IT2	Trucks with wheel diameter 645–900 mm	1.05	3.5	< 80%	exempt
IT3	Trucks with wheel diameter 900–2,000 mm	0.96	2.0	< 90%	< 2.0
IT4	All-terrain trucks	1.59	2.0	< 90%	< 2.0

Standards also exist for testing the vibration exposure of vehicle occupants. Some of these define artificial test tracks (e.g., tractor test code ISO 5008, 2002; industrial truck test code EN 13059, 2002). Although EN 1032 (2003) provides general guidance on methods of testing mobile machinery with the purpose of determining emission values, there is a lack of machine-specific guidance.

Most seat test codes require two types of trials to be completed in the laboratory. These are a test of the dynamic performance of the seat under simulated conditions similar to those experienced in the vehicle for which the seat is intended. The second test involves determination of the seat performance under worst-case conditions (i.e., sinusoidal motion at the seat resonance frequency). The test codes provide pass or fail criteria.

The original document should be consulted in preference to any summary when using a standard, as the summary will inevitably overlook some details. Additionally, standards are introduced, updated, or withdrawn constantly. Therefore, the user should take care to ensure that the current version of the standard is always used.

7 Hand-Transmitted Vibration Standards

7.1 INTRODUCTION

Many hand-transmitted vibration standards parallel those for whole-body vibration (see Chapter 6). They cover the similar broad areas of standards for measuring vibration exposure, vibration emission, and the performance of personal protective equipment (PPE). They are developed by the same bodies that set whole-body vibration standards, and the committees are often composed of the same individuals. Discussion of the development and position of standards found in Section 6.2 and Section 6.3 also apply for hand-transmitted vibration standards. Some aspects of hand-transmitted vibration standards are also covered elsewhere in this book (Chapter 4 and Chapter 5).

An organization with a potential hand-transmitted vibration problem could use standards in a variety of ways. Initially, they might require that the exposure of a worker operating a vibrating tool be assessed in an auditable way (Section 7.2). If the exposure of the worker was considered to be unacceptably high, then a replacement tool might be warranted, and declared vibration emission values might be used to help shortlist from a selection of possible options (Section 7.3). Finally, as a last option for protection, antivibration gloves might be considered as a form of PPE. These should be selected carefully, ensuring that the gloves pass the tests specified by the International Organization for Standardization (Section 7.4).

7.2 STANDARDS FOR HAND-TRANSMITTED VIBRATION MEASUREMENT AND ASSESSMENT

Between 1987 and 2001, there were two sets of standards providing guidance on the measurement and assessment of hand-transmitted vibration. These were International Standard ISO 5349 (1986) and British Standard BS 6842 (1987). In 2001, these were both replaced with the new version of ISO 5349. These three standards use similar techniques for measurement of vibration; therefore, it is usually possible to convert measurements made using one of the methods to an equivalent value that would have been obtained if using one of the other methods.

Although BS 6842 (1987) and ISO 5349 (1986) have been withdrawn, they are discussed in this section as there are many measurements in the literature that have been made using these techniques. Furthermore, this section will give some guidance on how to convert "old" measurements into equivalent values that would have been obtained if using the current ISO 5349 (2001).

7.2.1   ISO 5349-1: MECHANICAL VIBRATION — MEASUREMENT AND EVALUATION OF HUMAN EXPOSURE TO HAND-TRANSMITTED VIBRATION — PART 1: GENERAL REQUIREMENTS (2001)

ISO 5349-1 (2001) has also been adopted as a European standard (EN 5349-1, 2001) and as a national standard (e.g., BS 5349-1, 2001). It replaced ISO 5349 (1986), ENV 25349 (1992), and BS 6842 (1987). It is designed to specify general requirements for measuring and reporting hand-transmitted vibration exposures. The methods specified in the standard consider all three axes of vibration to which the hand is exposed. It does not provide a safe or unsafe vibration magnitude threshold. ISO 5349-1 makes frequent references to ISO 5349-2, which provides additional guidance on practical aspects of measuring hand-transmitted vibration (see Subsection 7.2.2).

At the heart of ISO 5349-1 is the requirement to measure triaxial vibration at the hand and to report the W_h frequency weighted r.m.s. acceleration magnitude. (Frequency weighting W_h is illustrated in Figure 4.3.) Ideally, the three axes of vibration should be measured simultaneously. The frequency weighted r.m.s. accelerations should be reported for each individual axis in addition to the vibration total value, a_{hv}, which is defined as the root-sum-of-squares of the individual measures made in the three orthogonal axes. If it is not possible to measure in all axes, then single axis measurements can be used to provide an estimated vibration total value, albeit with a large margin for error (70%). Further guidance on the mounting of accelerometers, duration of measurement, and estimation of vibration total values from single axis measurements is provided in ISO 5349-2.

From measurements of the vibration total value and estimates of the usage time of the tool (i.e., the tool "trigger time," sometimes referred to as the "anger time"), the daily exposure, normalized to an 8-h epoch, can be calculated. This is denoted as $A(8)$. The standard notes that it is difficult to obtain $A(8)$ values with an accuracy of better than 10%.

If measures are stated to be made in accordance with ISO 5349-1, then at least eight pieces of information should be reported (Table 7.1). These will not only allow the reader to understand what has been done but also to apply other standards to the same raw data.

There are six annexes to ISO 5349-1 (2001). One of these provides guidance on the health effects of hand-transmitted vibration, as well as a glossary of the types of disorders often described in the literature. Another annex provides a relationship between vibration exposures and health effects. This includes a dose–response relationship for the vibration exposures that predict a 10% prevalence of vibration-induced white finger (see Figure 4.4). Despite this relationship being included in the document, it is approximate; indeed, in the introduction, the standard indicates that one of the purposes of gathering occupational vibration exposure data is to "extend the present knowledge of dose–effect relationships."

TABLE 7.1

Minimum Information to Be Provided for Measurements of Hand-Transmitted Vibration [Made According to ISO 5349-1 (2001)]

Information to Be Reported

- The subject of the exposure evaluation
- The operations causing exposures to vibration
- The power tools, inserted tools, or workpieces involved
- The location and orientation of the transducers
- The individual root-mean-square, single axis frequency weighted accelerations measured
- The vibration total value for each operation
- The total daily duration for each operation
- The daily vibration exposure

The standard is careful not to specify safe and unsafe limits for hand-transmitted vibration. However, in a note in an annex it states:

Studies suggest that symptoms of the hand-arm vibration syndrome are rare in persons exposed with an 8-h energy-equivalent vibration total value, $A(8)$, at a surface in contact with the hand, of less than 2 m/s^2 and unreported for $A(8)$ values of less than 1 m/s^2.

These values are lower than the action value for hand-transmitted vibration (an $A(8)$ of 2.5 m/s^2) in the EU Physical Agents (Vibration) Directive (2002).

The squared relationship used in the calculation of the $A(8)$ produces a time dependency where a doubling of vibration magnitude would result in a reduction of exposure time by a factor of four to maintain the same vibration total value. Extremely high exposures are allowable if the exposure times are very short (Figure 7.1).

ISO 5349-1 (2001) is generally clear in the methodology that it specifies. The only part of the standard that is open to the judgment of the investigator is in the estimation of multiaxis measurements from single-axis vibration values. The Physical Agents (Vibration) Directive mandates the use of the standard while making risk assessments.

7.2.2 ISO 5349-2: MECHANICAL VIBRATION — MEASUREMENT AND EVALUATION OF HUMAN EXPOSURE TO HAND-TRANSMITTED VIBRATION — PART 2: PRACTICAL GUIDANCE FOR MEASUREMENT AT THE WORKPLACE (2001)

ISO 5349-2 (2001) has also been adopted as a European standard (EN 5349-2, 2001) and as a national standard (e.g., BS 5349-2, 2002). The aim of the standard is to give practical guidance on the implementation of ISO 5349-1 for real world measurements. It highlights precautions that should be taken and some potential pitfalls. It sets out, in the introduction, five distinct stages in evaluation of vibration exposure (Table 7.2).

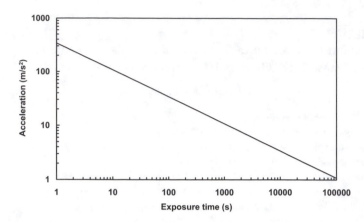

FIGURE 7.1 Relationship between acceleration magnitude and daily exposure time for a vibration total value $A(8)$ of 2 m/s^2 r.m.s., the threshold below which symptoms of hand-arm vibration syndrome are "rare" according to ISO 5349-1 (2001). A doubling of r.m.s. acceleration magnitude results in a reduction of exposure time by a factor of four for the same $A(8)$.

TABLE 7.2
The Five Distinct Stages in the Evaluation of a Vibration Exposure [According to ISO 5349-2 (2001)]

Stage	Description
1	Identifying a series of discrete operations that make up the subject's normal working pattern
2	Selection of operations to be measured
3	Measuring the r.m.s. acceleration value for each selected operation
4	Evaluation of the typical daily exposure time for each operation identified
5	Calculating the 8-h energy equivalent vibration total value (daily vibration exposure)

Whereas ISO 5349-1 requires a measurement at the hand, ISO 5349-2 advises on the selection of the measurement equipment, the location, and mounting method of the accelerometers and specifies how long each measurement should be carried out. Additional guidance is given on how to select the work operations to be measured.

According to ISO 5349-2, measurements of vibration should be made using at least three different workers, unless the data are required to be worker specific. It also states that the $A(8)$ should be calculated separately for both hands of the operators. Therefore, the assessment of an individual tool often requires at least six triaxial measurements.

Measurement uncertainty is quoted as often being 20 to 40%, primarily caused by the variability in the vibration inherent in the work operation and the precise mounting location for the accelerometers. Therefore, the investigator should take care to avoid unnecessary uncertainties, while acknowledging (and reporting) that a measurement uncertainty is not synonymous with measurement error.

ISO 5349-2 includes many examples throughout the document illustrating practical scenarios in which vibration measurements might be made. In particular, an informative annex provides five examples of approaches to vibration measurement for different types of situations (e.g., continuous tool vibration, intermittent tool vibration, vibration from a variety of tools).

Another annex gives examples of measurement locations for 39 different types of power tools. Most of these measurement locations are drawn from the type test standards such as ISO 8662 (see Subsection 7.3.2), ISO 7505, and ISO 7916. Some of these are illustrated in Figure 5.16.

ISO 5349-2 (2001) and its partner, ISO 5349-1 (2001), are an extremely useful pair of documents. They are relatively clear and provide the reader with enough information to make basic vibration measurements in most occupational environments. The standards do not attempt to provide exhaustive guidance on more complex methods, such as frequency analysis, although this is often a useful addition to any vibration evaluation. As such, they are appropriate for newcomers to the field of human vibration and should empower them with the knowledge that they need to make basic measurements.

7.2.3 BS 6842: British Standards Guide to Measurement and Evaluation of Human Exposure to Vibration Transmitted to the Hand (Superseded) (1987)

BS 6842 (1987) replaced the BSI Draft for Development DD 43 (1975) and was itself superseded by BS 5349-1 in 2001 (i.e., ISO 5349-1, 2001). In most respects, BS 6842 and ISO 5349-1 (2001) are compatible. The same W_h frequency weighting is used for both standards and both normalize vibration exposures to an 8-h $A(8)$ value.

The most substantial difference between BS 6842 and ISO 5349-1 is that the British standard was based on measurements of vibration in the dominant axis, rather than the combined axes of the later standard. Although it is not possible to provide a generalized conversion from a single "dominant" axis measurement to an equivalent measurement made over orthogonal axes, some general principles can be applied. In many cases, vibration could have been measured in all axes to determine which was the dominant axis. If these data are reported, then the values can be combined in accordance with ISO 5349.

If there is no information on the relative contribution from the three individual axes, it might be possible to scale the dominant axis measurement, made according to BS 6842, to estimate what the measurement would have been according to ISO 5349. If there was no vibration in the nondominant axes, and the magnitude in the dominant axis was a_{dom}, then the root-sum-of-squares expression to calculate the vibration total value (a_{total}) would be:

$$a_{total} = \sqrt{a_{dom}^2 + 0.0^2 + 0.0^2} = a_{dom}$$

Thus, in the best case, the measurements made according to the two standards would be identical. If the vibrations in each of the axes were identical to that in the dominant axis, then the vibration total value would be:

TABLE 7.3
Ratio of Root-Sum-of-Squares Vibration Magnitude to the Dominant Axis for Six Categories of Vibrating Tools [Data from Nelson (1997)]

Category	Example Tools	Mean Ratio	Standard Deviation
Grinders	Angle grinder, straight grinder	1.32	0.15
Sanders and polishers	Rotary sander, orbital sander	1.34	0.19
Scaling tools	Scaling hammer, needle scaler	1.17	0.12
Miscellaneous percussive tools	Chipping hammer, pneumatic chisel	1.26	0.13
Miscellaneous rotating and reciprocating tools	Power saw, pistol drill	1.37	0.17
Workpieces	Pedestal grinder	1.28	0.17

$$a_{total} = \sqrt{a_{dom}^2 + a_{dom}^2 + a_{dom}^2} = 1.73 \times a_{dom}$$

Thus, in the worst case, the measurements made according to ISO 5349 would be 73% higher than those made according to BS 6842.

Nelson (1997) compares the root-sum-of-squares with the dominant axis vibration for 113 vibration measurements from a variety of tools. The range encompassed values for individual tools that had emissions close to the theoretical best case and worst case. However, the ratio of root-sum-of-squares to dominant axis for most tools was between 1.2 and 1.4. Scaling tools had the lowest mean ratio (1.17); miscellaneous rotating and reciprocating tools (e.g., drills and saws) had the largest mean ratio (1.37) (Table 7.3). Nelson proposes a multiplying factor of 1.4 to be used when converting from single axis measurements to estimating the root-sum-of-squares. Considering the measurement uncertainty with these types of measures, this value seems reasonable.

7.2.4   ISO 5349: MECHANICAL VIBRATION — GUIDELINES FOR THE MEASUREMENT AND THE ASSESSMENT OF HUMAN EXPOSURE TO HAND-TRANSMITTED VIBRATION (SUPERSEDED) (1986)

ISO 5349 (1986) was superseded by ISO 5349-1 in 2001. The two standards are generally compatible as they use the same general methods, axes, and frequency weighting. However, there are two important differences. First, the old version of the standard was based on measurements of vibration in the dominant axis rather than the root-sum-of-squares as specified in the new version of ISO 5349. The nature of this dissimilarity is identical to the difference between BS 6842 (1987) and ISO 5349 (2001), and therefore the methods for addressing the problem of converting between standards suggested in the previous section also apply here.

The second difference between the 1986 and 2001 versions of ISO 5349 is that the old version used a reference period of 4 h rather than the 8-h reference period used in the new version. The conversion between these reference periods is straight-forward:

$$A(8) = \frac{A(4)}{\sqrt{2}}$$

or

$$A(8) \approx 0.7 \times A(4)$$

where $A(8)$ and $A(4)$ are the 8-h and 4-h energy equivalent frequency weighted vibration magnitudes, respectively.

7.3 STANDARDS FOR DETERMINING THE VIBRATION EMISSION VALUE OF TOOLS AND HAND-GUIDED MACHINES

The situation faced by purchasers of tools that expose the hands to vibration is similar to that of the purchasers of mobile machinery (see Section 6.6). Risks from vibration exposure must be minimized, and one way of doing this is by purchasing tools with a low vibration emission. The EU Machinery Safety Directive requires that measurements of vibration be made by the manufacturer and that the results are available to the purchaser. Therefore, it should be possible for a buyer to select a low-vibration tool in preference to other tools, should they so wish. As the vibration emission of a tool is affected by the manner of operation, all manufacturers should produce emission values, using similar techniques. Hence, there is a need for test codes by which tools can be evaluated.

The most well known set of test codes for determining the vibration emission of tools is the ISO 8662 series, although there are many other standards serving a similar purpose for other types of tools such as chainsaws (ISO 7505, 1986) and brush saws (ISO 7916, 1989). This section does not aim to provide a comprehensive précis of all current type test standards, but to describe the general approach common to most, while highlighting some aspects of the standards that can be troublesome and using others as examples.

7.3.1 EN 1033: HAND-ARM VIBRATION — LABORATORY MEASUREMENT OF VIBRATION AT THE GRIP SURFACE OF HAND-GUIDED MACHINERY — GENERAL (1995)

EN 1033 (1995) has also been adopted as a national standard (e.g., BS 1033, 1996). It is designed to allow test codes to be developed such that the vibration emission from different machinery within a class can be compared. The standard provides general guidance for the laboratory assessment of all types of hand-guided machinery, rather than specific procedures for any class. Hand-guided machinery includes machines such as lawn mowers and vibratory rollers. The standard does not apply to handheld power tools, situations in which vibration is transmitted through a workpiece (e.g., pedestal grinding), or for steering wheels. There are similarities between EN 1033 and ISO 8662-1 (1988); many paragraphs are identical in both standards.

The key measurement to be made according to EN 1033 is the frequency-weighted r.m.s., measured in the dominant axis of vibration. If there is no clear dominant axis (defined as an axis in which vibration magnitude is more than double that in the other two axes), then triaxial measurements should be made. Accelerometers should be mounted at the midpoint of the position where the operator holds the tool. Three operators should be used for the tests, and the emission value is the arithmetic mean of all tests. Occasionally, additional information might be required, such as details of the frequency content in the vibration signal and the grip force, feed force, and speed setting of the machinery.

One area where EN 1033 is unable to provide guidance is the detailed specification of the working procedure to be measured for any test code based on the standard. Ideally, the working procedure should be typical for the tool usage, but this depends on the vibration measurements being reproducible. If there is an unacceptably large variation in the vibration emission depending on the task, then artificial procedures can be defined. Therefore, some test codes developed from EN 1033 might specify realistic conditions for the test, whereas others might specify artificial conditions.

The premise of trying to define test codes such that the variability in the vibration emission is minimized must be questioned. If there is a large range of possible vibration magnitudes for a tool in use, then stating this in the sales literature, by referring to the mean and standard deviation for a variety of test conditions, could be more beneficial than giving a single value that will rarely be experienced. Reporting the uncertainty would clearly indicate where ranking of tools by emission value is inappropriate. Furthermore, it would highlight to the purchaser the inexact nature of vibration exposures and could encourage testing in the real work environment for each individual tool, which is the ideal situation for making a risk assessment.

7.3.2 ISO 8662: Hand-held Portable Power Tools — Measurements of Vibrations at the Handle — Part 1: General (1988)

ISO 8662-1 (1988) has also been adopted as a European standard (EN 28662-1, 1992) and as a national standard (e.g., BS 28662-1, 1993). The standard is, in many respects, similar to EN 1033 (1995) in that it forms a basis of a measurement technique to establish emission values for a category of device (i.e., portable power tools). It defines a laboratory measuring procedure that should be as realistic and reproducible as possible (although this is not always achieved; see Subsection

TABLE 7.4
Parts of International Standard ISO 8662, Hand-held Portable Power Tools —
Measurement of Vibrations at the Handle

Standard Reference	Publication Year	Part	Subtitle
ISO 8662-1	1988	Part 1	General
ISO 8662-2	1992	Part 2	Chipping hammers and riveting hammers
ISO 8662-3	1992	Part 3	Rock drills and rotary hammers
ISO 8662-4	1994	Part 4	Grinders
ISO 8662-5	1992	Part 5	Pavement breakers and hammers for construction work
ISO 8662-6	1994	Part 6	Impact drills
ISO 8662-7	1997	Part 7	Wrenches, screwdrivers, and nut runners with impact, impulse, or ratchet action
ISO 8662-8	1997	Part 8	Polishers and rotary, orbital, and random orbital sanders
ISO 8662-9	1996	Part 9	Rammers
ISO 8662-10	1998	Part 10	Nibblers and shears
ISO 8662-11	1999	Part 11	Fastener driving tools
ISO 8662-12	1997	Part 12	Saws and files with reciprocating action and saws with oscillating or rotating action
ISO 8662-13	1997	Part 13	Die grinders
ISO 8662-14	1996	Part 14	Stone-working tools and needle scalers

7.3.1 for a discussion of the possible drawbacks of creating artificially repeatable test codes).

Vibration measurements should be made at the handle of tools and expressed as the r.m.s., although the magnitude might also be expressed in terms of decibel acceleration, L_{a_h} :

$$ L_{a_h} = 20 \log\left(\frac{a_h}{a_0} \right) $$

where a_h is the r.m.s. acceleration, and a_0 is the reference acceleration (10^{-6} m/s² r.m.s.). As for EN 1033, measurements are usually made in the dominant axis only, although it is possible to standardize to measuring all three axes. Only one operator is required, according to ISO 8662-1.

There are 14 parts to ISO 8662 encompassing a wide range of power tools (Table 7.4). All of the parts are similar in approach (based on ISO 8662-1) but give specific guidance on the details of each test. Parts 2, 4, and 6 are covered here as illustrations of parts of the standard that use artificial tests, parts that need to be interpreted with caution, and parts that use realistic tests, respectively.

7.3.3 ISO 8662: Hand-held Portable Power Tools —
Measurements of Vibrations at the Handle — Part 2:
Chipping Hammers and Riveting Hammers (1992)

ISO 8662-2 (1992) has also been implemented as a European standard (EN 28662-2, 1994) and as a national standard (e.g., BS 28662-2, 1995). The standard specifies an artificial test procedure that uses an "energy absorber." (A similar device is used in Part 3 of ISO 8662 for assessment of rock drills and Part 5 of ISO 8662 for assessment of pavement breakers and pick hammers.) The energy absorber consists of a steel cylinder partly filled with 4-mm-diameter steel balls (Figure 7.2). Loading a percussive tool with a steel ball energy absorber should approximate a normal working situation.

Tests are carried out with the operator standing on a scale so that the pressure applied on the tool can be monitored. Operators should apply a constant feed force

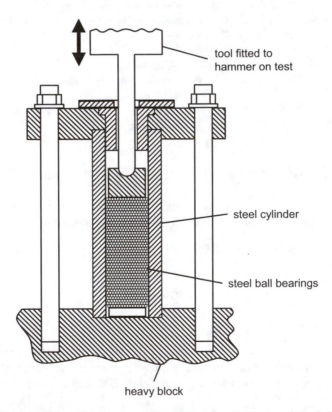

FIGURE 7.2 Design of steel ball energy absorber for standardized testing of chipping hammers and riveting hammers. [Adapted from International Organization for Standardization (1992). Hand-held portable power tools: measurement of vibrations at the handle — part 2: chipping hammers and riveting hammers. ISO 8662-2. Geneva: International Organization for Standardization.]

of approximately 40 times the mass of the tool. Three operators should be used, and tests continued until five consecutive frequency-weighted r.m.s. measurements obtained using the same operator have a coefficient of variation (i.e., the ratio of the standard deviation to the mean) of less than 0.15. Vibration is only measured in the direction of action of the tool.

7.3.4 ISO 8662: HAND-HELD PORTABLE POWER TOOLS — MEASUREMENTS OF VIBRATIONS AT THE HANDLE — PART 4: GRINDERS (1994)

ISO 8662-4 (1994) has also been implemented as a European standard (EN 8662-4, 1995) and as a national standard (e.g., BS 8662-4, 1995). The grinder test code is one of the most controversial standards and has been the focus of a body of research (e.g., Ward, 1996; Stayner, 1997; Smeatham, 2000). There are three sources of vibration for a grinder: the motor and drive, the reaction forces from the grinding wheel on the workpiece, and any vibrations caused by unbalanced grinding wheels rotating at high speed (as grinding wheels wear down, the extent of the unbalance constantly changes). The motor and drive do not substantially contribute to the vibration, but the reaction forces from the grinding task and unbalance forces are the main contributors to the vibration at the handles. ISO 8662-4 only considers the vibration from unbalanced grinding wheels and neglects any contribution from the grinding task itself.

Assessments made according to ISO 8662-4 measure the vibration at both handles of the grinder. The grinding wheel is replaced by an aluminum disc with an offset aperture to produce a known unbalance (Figure 7.3). The grinder is tested with three skilled operators who apply a known feed force (that is specified according to the grinding disc diameter), by counteracting a load suspended on a pulley system (Figure 7.4). Repeated measurements are made with the grinder running free (i.e., suspended in the air with no contact with a workpiece). At no stage during the test procedure does the grinder grind.

In order to assess the applicability of ISO 8662-4, Ward (1996) reports vibration emissions of 10 grinders. The manufacturers' declared values, repeat tests of declared values, and field measurements were compared (Figure 7.5). These data showed that for three of the grinders (A, C, and D) the declared values were substantially lower than those obtained when retested in the laboratory. The two tools with the lowest declared values also had the lowest mean vibration emissions measured in the field; the tool with the highest declared value had the highest mean vibration emission measured in the field. The rank order of the other tools in terms of field vibration magnitudes, however, did not follow the pattern that would be suggested from the declared values; for example, tool C had the second highest mean field vibration measurement but was the third lowest in terms of declaration. Generally, field measurements were higher than the declared values. These data illustrate that using declared values to estimate vibration exposures should be used with caution, at least for grinders.

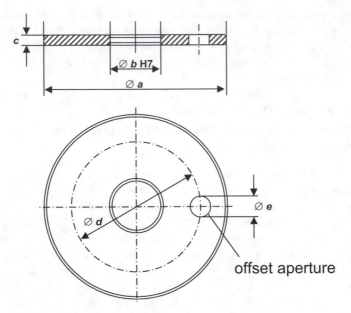

FIGURE 7.3 Specification of aluminum test wheel with offset aperture to be used when type-testing grinders. Dimensions depend on the size of the grinder. [Adapted from International Organization for Standardization (1994). Hand-held portable power tools: measurement of vibrations at the handle — part 4: grinders. ISO 8662-4. Geneva: International Organization for Standardization.]

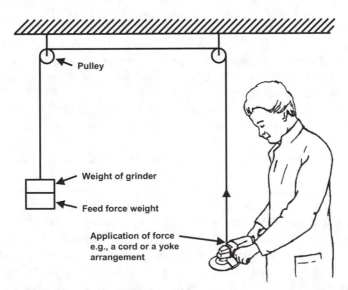

FIGURE 7.4 Working position of skilled operator while testing a grinder according to ISO 8662-4 (1994). The feed force weight depends on the diameter of the grinding wheel. [From British Standards Institution. (1995). Hand-held portable power tools: measurement of vibrations at the handle — part 4: grinding machines. BS 8662-4. London: British Standards Institution.]

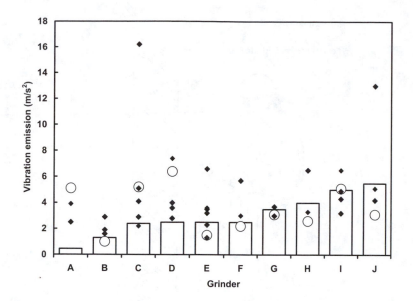

FIGURE 7.5 Comparison of declared values (bars), retested vibration emission levels according to ISO 8662-4 (circles), and field measurements (diamonds) for 10 different grinders. [Data from Ward (1996), ranked according to declared value.]

7.3.5 INTERNATIONAL STANDARD ISO 8662: HAND-HELD PORTABLE POWER TOOLS — MEASUREMENTS OF VIBRATIONS AT THE HANDLE — PART 6: IMPACT DRILLS (1994)

ISO 8662-6 (1994) has also been implemented as a European standard (EN 8662-6, 1995) and as a national standard (e.g., BS 8662-6, 1995). The standard specifies a test that is relatively realistic — the impact drill is used to drill into concrete. This is in contrast to many other parts of ISO 8662 where the tool is not performing the task for which it was designed.

Test subjects must stand on a platform that can measure the operator's feed force, which must be maintained between 150 and 180 N. The operators must drill into a concrete wall using a new 8-mm drill bit (Figure 7.6). The measurements should commence once the drill bit has reached a depth of 10 mm, so that the initial starting of the hole is excluded from the measurement. Three skilled operators are required for the tests, which are continued until five consecutive frequency-weighted r.m.s. measurements obtained using the same operator have a coefficient of variation of less than 0.15.

7.4 STANDARDS FOR TESTING THE DYNAMIC PERFORMANCE OF ANTIVIBRATION GLOVES

Antivibration gloves are attractive as a form of personal protective equipment (see also Subsection 4.5.5.3). A Utopian glove would provide a cost-effective solution for substantially reducing the vibration at the hand, while not increasing other risk

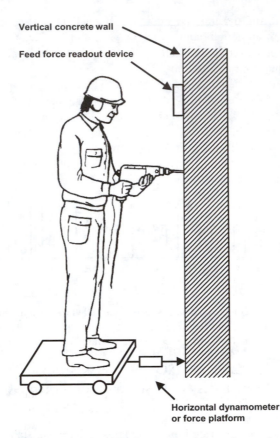

FIGURE 7.6 Working position of a skilled operator while testing an impact drill according to BS EN ISO 8662-6 (1995).

factors (e.g., not increasing, grip forces). Despite recent improvements in their design, antivibration gloves do not currently offer this idealistic solution, although it is possible that developments in materials technology could provide better gloves in the future.

There are two approaches to testing antivibration gloves. One is to test the material from which the glove is made, while loaded (see description in following subsection) with a simulation of the human hand-arm system (e.g., ISO 13753, 1998; Japanese Industrial Standard JIS T 8114, 1987); the second approach is to test a complete glove worn by an operator (e.g., ISO 10819, 1996).

7.4.1 ISO 13753: Mechanical Vibration and Shock — Hand-Arm Vibration — Method for Measuring the Vibration Transmissibility of Resilient Materials when Loaded by the Hand-Arm System (1998)

ISO 13753 (1998) has also been implemented as a European standard (EN 13753, 1998) and as a national standard (e.g., BS 13753, 1999). The standard describes a method for testing the dynamic performance of resilient materials that might be intended for use in an antivibration glove or on a tool handle. The standard does not provide pass or fail criteria but allows for materials (and layered materials) to be ranked according to transmissibility.

A circular piece of the material to be tested (with a diameter of at least 90 mm) is loaded with a 2.5-kg mass on a shaker table. The sample is exposed to vibration between 10 and 500 Hz while the vibration on the loading mass and the shaker table are measured. The ratio of the accelerations provides the (complex) transmissibility of the sample while loaded with the mass. From the measurements of transmissibility, the material's mechanical impedance can be calculated. Combining the material impedance with a standardized hand-arm impedance (as defined in the standard) allows for the transmissibility of the material to be calculated while loaded with a human hand-arm system.

This technique assumes that the dynamics of the material are linear and that the impedance of the hand-arm system is linear and adequately represented by the model in the standard.

7.4.2 ISO 10819: Mechanical Vibration and Shock — Hand-Arm Vibration — Method for the Measurement and Evaluation of the Vibration Transmissibility of Gloves at the Palm of the Hand (1996)

ISO 10819 (1996) has also been implemented as a European standard (EN 10819, 1996) and as a national standard (e.g., BS 10819, 1997). The standard defines a test for gloves to ensure that their dynamic performance meets minimum standards if they are to be sold as "antivibration" gloves. The standard was developed in the context of there being no gloves that would pass. The standard states:

> Within the current state of knowledge, gloves do not provide sufficient attenuation in the frequency range below 150 Hz.

Some studies using the standard have found that no gloves pass (e.g., Paddan and Griffin, 1999), although more recent reports have shown some gloves passing the test (e.g., Hewitt, 2002). One could interpret this trend as positive, because this shows that glove manufacturers have been able to meet the target set in ISO 10819 (see also Subsection 4.5.5.3).

The glove tests are carried out in the palm of the hand. There is no measurement required at the fingers (although this is the location where vibration white finger occurs). Nevertheless, the standard requires that the vibration-isolating material extend over the fingers.

ISO 10819 (1996) requires that three sets of measurements are made of the dynamic performance of the glove being tested, each set being carried out while the glove is worn by a different subject. The tests are carried out in the laboratory, using a shaker fitted with a specially designed handle capable of measuring the subjects' grip force (which should be 25 to 35 N) and push force (which should be 42 to 58 N). Acceleration is measured at two locations: on the surface of the handle and

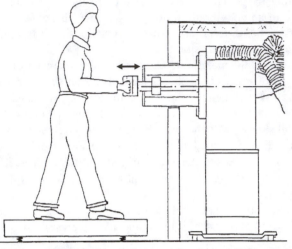

Operator on adjustable platform   Vibration excitation system

FIGURE 7.7 Posture of operator while testing antivibration gloves according to BS EN ISO 10819 (1997).

inside the glove, using an accelerometer mounted within a moulded "palm adapter" that is gripped between the hand and glove while holding the vibrating handle. The hand is orientated vertically and vibrates in the hand-z direction (Figure 7.7).

Measurements of the r.m.s. acceleration magnitude are made using two vibration spectra, termed the M (medium) spectrum and the H (high) spectrum, which contain most energy from 31.5 to 200 Hz and from 200 to 1000 Hz, respectively (Figure 7.8). The M and H spectra must be generated at a W_h frequency-weighted magnitude of 3.4 and 3.3 m/s^2 r.m.s. (16.7 and 92.2 m/s^2 r.m.s. unweighted), respectively. Measurements must also be made using only the palm adapter gripped against the surface of the handle ("bare hand" condition), to use as a reference to correct the calculations of glove transmissibility, should there be any discrepancy between the calibration of the two accelerometers. Each glove should be tested twice for each of the three subjects.

The transmissibility is calculated as the ratio of the frequency-weighted r.m.s. acceleration measured at the palm to the same type of acceleration measured on the handle (corrected for discrepancies in calibration). Mean transmissibilities are calculated for the two repeat measurements with the three subjects for both of the vibration spectra and expressed as $\overline{TR}_M$ and $\overline{TR}_H$ for the M and H spectra, respectively. Note that these measurements are not the same as transmissibilities measured in the frequency domain as described in Subsection 5.8.3.3, but are analogous to r.m.s. SEAT values (see Subsection 2.5.2), albeit measured with artificial stimuli.

For a glove to be classed as an "antivibration glove" according to ISO 10819 all of three criteria must be met. First, for the tests using the M spectrum:

$$\overline{TR}_M < 1,$$

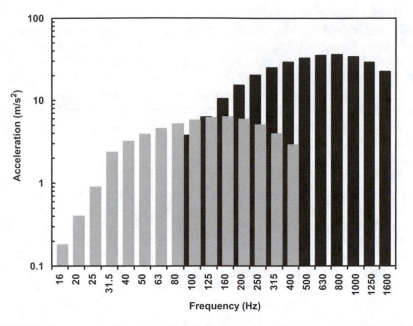

FIGURE 7.8 Test spectra M (grey bars) and H (black bars) for testing the transmissibility of antivibration gloves according to ISO 10819 (1996), expressed as one third-octave values.

i.e., the glove must, on average, attenuate in the frequency range of 31.5 to 200 Hz. Second, for the tests using the H spectrum:

$$\overline{TR}_H < 0.6,$$

i.e., the glove must, on average, transmit less than 60% of the vibration in the frequency range of 200 to 1000 Hz. The third criterion is that the glove material at the palm must extend over the fingers.

One problem with using the criteria based on averaged transmissibility values is that the frequency dependence of the dynamic response of the glove is unclear to the purchaser. It is likely that the glove will amplify the vibration at some frequencies within the M spectrum, although to pass the test it must attenuate at others (Paddan and Griffin, 1999). Therefore, if the vibration frequency coincides with the glove resonance frequency, then the glove will not reduce the vibration, but amplify it (see also Subsection 4.5.5.3).

7.5 CHAPTER SUMMARY

Hand-transmitted vibration standards can be categorized in three ways: those that can be used for measuring and assessing exposure in the workplace, those for determining vibration emission of tools and machinery, and those for testing the dynamic performance of antivibration gloves.

The two parts of ISO 5349 should be used to assess vibration in the field. These require that triaxial vibration be measured at the handle of the tool or machine and W_h frequency weighted. The vibration total value is the root-sum-of-squares of the measurements made in the three orthogonal axes. Combining the vibration total value with the tool usage time, the $A(8)$ can be calculated. ISO 5349 does not provide clear exposure limits, but it does include an approximate dose–response relationship that predicts a 10% prevalence of vibration-induced white finger from vibration magnitude and number of years of exposure. Earlier, and now superseded, standards (BS 6841, 1987; ISO 5349, 1986) only require measurement of vibration in the dominant axis.

There are many standards that define vibration emission test codes for tools and hand-guided machines. The most extensive series are the 14 parts of ISO 8662 for the assessment of handheld portable power tools. Depending on the specific tool type, tests are carried out under artificial or realistic conditions. For example, chipping hammers, pavement breakers, and rock drills are tested using an artificial "energy absorber"; grinders are measured while free running but fitted with an aluminum disc of known unbalance; impact drills are tested while drilling into concrete. These type tests are used to determine the vibration emission so that the declared values required by the EU Machinery Directive can be reported in the tool literature.

ISO 10819 (1996) defines tests that must be passed for a glove to be classed as an "antivibration glove." The glove is tested using two vibration spectra (the M and H spectra). The ratio of acceleration at the palm of the hand to that at the handle for the two test spectra are used as the basis for passing (or failing) the test.

As for all standards, the original document should always be consulted in preference to any summary. Standards are introduced, updated, or withdrawn constantly, and therefore the user should take care to ensure that the current version of any standard is always used.

8 European Directives

8.1 INTRODUCTION

Among the most important documents relating to human response to vibration are two European Directives. These Directives regulate the reporting of the vibration emission and design principles for machines and provide limits on the vibration that any worker can be exposed to. Their impact is, of course, greater within Europe than throughout the rest of the world, but any machine intended to be sold globally will need to meet the demands of the European market. Furthermore, if these Directives prove successful within Europe, other nations are likely to introduce similar domestic legislation.

Many pieces of legislation in force across Europe are derived from European Directives. For example, European member-state national regulations relating to manual handling (e.g., the U.K. Manual Handling Operations Regulations; Her Majesty's Stationery Office, 1992) are an implementation of the European Directive 90/269/EEC (European Commission, 1990). This means that the legislation is similar across Europe, thereby ensuring that workers in each state are equally protected, as well as ensuring that industries in one state cannot cut costs by reducing safety standards.

European Directives are developed through a legal process by the European Union (EU) (Section 8.2) and must be implemented in member states (Section 8.3). Within the context of human response to vibration, the most important Directives are the Machinery Directive (Section 8.3) and Physical Agents (Vibration) Directive (Section 8.4). Those responsible for occupational health in the workplace may also be affected by the Physical Agents (Noise) Directive (briefly discussed in Section 8.5).

8.2 DEVELOPMENT OF EUROPEAN DIRECTIVES

The procedure for legislation within the EU involves three institutions: the European Parliament, the Council of the European Union, and the European Commission. The European Parliament comprises 626 members (MEPs) directly elected by citizens of member states and who work in the interest of their constituents; the Council of the European Union is made up of representatives of member-state governments and works in the interest of the member states; the European Commission is an independent body responsible for drafting legislation and consists of 20 members approved by the European Parliament and the governments of the EU member states.

The "codecision" procedure for developing European legislation entails the Commission drafting the legislation for acceptance by both the Parliament and the Council. After the initial drafting of the document by the Commission, the document has its first reading in the Parliament, and an opinion is issued. Taking into consideration the Parliament's opinion, the Council modifies the legislation to reach a "common position." The process of reaching a common position is somewhat iterative, and it can take many drafts of a potential Directive before agreement is achieved. Once a common position is reached, both institutions have, in principle, accepted that the legislation should be adopted, although the details still require work. A second round of modifications takes place through the Parliament and Council after which, if full agreement is still not reached, a Conciliation Committee is convened with representatives from both Parliament and Council, who must work together. This committee must reach a full agreement within 6 weeks, and this text is either confirmed or rejected (with no further opportunity for amendment) by the Council and Parliament. The agreed text is published in the *Official Journal of the European Communities* that contains details of when the Directive must be complied with.

8.3   LEGAL POSITION OF EUROPEAN DIRECTIVES

The European Directives that relate to human response to vibration are mandatory within the EU (Austria, Belgium, Cyprus, Czech Republic, Denmark, Estonia, Finland, France, Germany, Greece, Hungary, Ireland, Italy, Latvia, Lithuania, Luxembourg, Malta, the Netherlands, Poland, Portugal, Spain, Slovakia, Slovenia, Sweden, and the UK) and the additional states that complete the European Economic Area (Iceland, Liechtenstein, and Norway).

Directives are legal documents that must be complied with; offending member states can be penalized by the European Commission. Often, the Directives refer to international or European standards, thereby giving the standard a legal framework within the affected countries. Directives must be implemented within domestic law by a date specified in the text.

Some Directives [e.g., the Physical Agents (Vibration) Directive] allow slightly alternative forms, depending on the preference of the member state. Some can also have a phased implementation according to the industrial sector. Such an introduction can occur if there are some sectors that find it more difficult or expensive to comply than others. Also, these often have a powerful political lobby.

Directives are enforced at a domestic level. There is no central European inspectorate. Although all member states must implement adopted Directives and provide reports to the Commission regarding practical implementation, the resources available to enforce compliance vary across the EU. This is a problem of enforcement rather than a problem with the Directives, *per se*.

8.4   THE MACHINERY SAFETY DIRECTIVE: DIRECTIVE 98/37/EC OF THE EUROPEAN PARLIAMENT AND OF THE COUNCIL ON THE APPROXIMATION OF THE LAWS OF THE MEMBER STATES, RELATING TO MACHINERY

The Machinery Safety Directive of June 22, 1998 (European Commission, 1998) consolidated the earlier version of the Directive of June 14, 1989 (Council Directive

TABLE 8.1
Machinery That Might Expose the Operator to Vibration, but Are Specifically Excluded from the Scope of the Machinery Safety Directive

- Machinery whose only power source is directly applied manual effort, unless it is a machine used for lifting or lowering loads
- Machinery for medical use used in direct contact with patients
- Special equipment for use in fairgrounds and/or amusement parks
- Machinery specially designed or put into service for nuclear purposes which, in the event of failure, may result in an emission of radioactivity
- Firearms
- Means of transport, i.e., vehicles and their trailers intended for transporting passengers and/or goods by air or on road, rail, or water networks. Vehicles used in the mineral extraction industry are not excluded
- Seagoing vessels and mobile offshore units together with equipment on board such vessels or units
- Cableways, including funicular railways
- Agricultural and forestry tractors (as defined in Directive 74/150/EEC)
- Machines specially designed and constructed for military or police purposes
- Lifts, elevators, winding gear, and hoists intended for lifting persons
- Means of transport of persons using rack-and-pinion rail-mounted vehicles

89/392/EEC, European Commission, 1989) and its many amendments. For the purposes of the Directive, "machinery" means " … an assembly of linked parts or components, at least one of which moves … in particular for the processing, treatment, moving, or packaging of a material."

However, there are many categories of "machinery" that are excluded from the Directive (Table 8.1), including agricultural and forestry tractors, various means of transport, and medical equipment (e.g., surgical power tools).

8.4.1 PRINCIPLES OF SAFETY INTEGRATION

The Machinery Safety Directive is designed to set mandatory minimum health and safety requirements. The principles of safe design of machines must be incorporated into all foreseeable uses of the machine, including its assembly and dismantling, as well as into predictable abnormal uses that might be beyond the designed scope of application for the machine. The Directive states that the manufacturer must apply principles of safety integration into the design in the following order:

1. Eliminate or reduce risks as far as possible (i.e., with inherently safe machinery design and construction).
2. Take the necessary protection measures in relation to risks that cannot be eliminated.
3. Inform users of the residual risks due to any shortcomings of the protection measures adopted, indicate whether any particular training is required, and specify any need to provide personal protection equipment.

TABLE 8.2
Summary of Human Vibration and Noise
Information To Be Provided in the Instruction Book
According to the Machinery Safety Directive (1998)

Hand-Transmitted Vibration

< 2.5 m/s² r.m.s.	Indicate threshold is not exceeded
> 2.5 m/s² r.m.s.	Report emission value

Whole-Body Vibration

< 0.5 m/s² r.m.s.	Indicate threshold is not exceeded
> 0.5 m/s² r.m.s.	Report emission value

Noise

< 70 dB(A) L_{eq}	Indicate threshold is not exceeded
> 70 dB(A) L_{eq}	Report emission value
> 85 dB(A) L_{eq}	Report emission value and sound power level
> 130 dB(C) peak	Report peak pressure

The Directive specifically refers to vibration and states that:

Machinery must be so designed and constructed that risks resulting from vibrations produced by the machinery are reduced to the lowest level, taking account of technical progress and the availability of means of reducing vibration, in particular, at source.

This is expanded in many sections of the Directive, including those relating to the driving position and the design of the seat (which must be designed to reduce vibration transmitted to the driver to the lowest level that can be reasonably achieved).

8.4.2 REPORTING OF EMISSION VALUES FOR HAND-TRANSMITTED VIBRATION

The third of the design principles listed in the preceding section (informing users of residual risks) forms the basis for mandating some parts of the information that must be included in the instruction book (Table 8.2). The Directive states that:

The instructions must give the following information concerning vibrations transmitted by hand-held and hand-guided machinery:

• The weighted root mean square acceleration value to which the arms are subjected, if it exceeds 2.5 m/s² as determined by the appropriate test code. Where the acceleration does not exceed 2.5 m/s², this must be mentioned.

If there is no appropriate test code, the manufacturer must indicate the measurement methods and conditions under which measurements were made."

This requirement for test codes was one of the main motivators for the development of the codes described in Section 7.3 (e.g., the ISO 8662 series of standards).

8.4.3 REPORTING OF EMISSION VALUES FOR WHOLE-BODY VIBRATION

For whole-body vibration emission, the Directive states that:

... the instruction handbook must contain the following information:

a. regarding the vibrations emitted by the machinery, either the actual value or a figure calculated from measurements performed on identical machinery:
  • ... the weighted root mean square acceleration value to which the body (feet or posterior) is subjected, if it exceeds 0.5 m/s^2, should it not exceed 0.5 m/s^2, this must be mentioned.

If an appropriate test code exists, it should be used (see Section 6.6); otherwise, the manufacturer must indicate the operating conditions of the machinery during measurement and which methods were used for taking measurements. One of the difficulties in defining test codes for whole-body vibration is that many machines are used with several forms of attachment (often sourced from different manufacturers) and in widely varying terrain. In this case, the manufacturers of both parts of the machine have a responsibility to provide information to enable the equipment to be operated safely.

8.4.4 IMPACT OF THE MACHINERY SAFETY DIRECTIVE ON OCCUPATIONAL HEALTH OF WORKERS

The Machinery Safety Directive in itself should provide some protection for workers operating machines that have been bought within Europe since its enforcement in its original form on January 1, 1993. The requirement that vibration be reduced as far as possible for every machine means that all new machines should be of "low vibration." Nevertheless, the most tangible improvement in the working environment (as far as vibration is concerned) can be achieved by selecting tools and machines with the lowest vibration emission (assuming that the declared values are representative of the vibration exposures experienced by the operator throughout the life of the tool). Freely accessible electronic databases are available that collate these declared values and should assist in short-listing tools with low declared values. There are difficulties in standardizing test codes for whole-body vibration (see Subsection 6.6.1), and there are doubts over the validity of using declared values as a basis for selecting power tools (see Subsection 7.3.1). However, until "perfect" test codes can be developed, purchasers are unlikely to have a better alternative for short-listing tools.

8.4.5 REPORTING OF EMISSION VALUES FOR NOISE EXPOSURE

Analogous emission declaration requirements for noise exposures are included in the Directive. With regard to providing information in the instructions, the Directive requires the reporting of equivalent, continuous A-weighted sound pressure levels at workstations where they exceed 70 dB(A) [as well as where the level does not exceed 70 dB(A)]; peak C-weighted instantaneous sound pressure levels where they exceed 130 dB(C) peak; and the sound power level where the equivalent continuous A-weighted sound pressure level at workstations exceeds 85 dB(A). Further comment on the noise aspects of the Directive is beyond the scope of this book.

8.5 THE PHYSICAL AGENTS (VIBRATION) DIRECTIVE: DIRECTIVE 2002/44/EC OF THE EUROPEAN PARLIAMENT AND OF THE COUNCIL ON THE MINIMUM HEALTH AND SAFETY REQUIREMENTS REGARDING THE EXPOSURE OF WORKERS TO THE RISKS ARISING FROM PHYSICAL AGENTS (VIBRATION)

The Physical Agents (Vibration) Directive of June 25, 2002 (European Commission, 2002) was the first of an intended series of directives specifying limits on exposure to physical agents. This stemmed from the European Parliament's instigation of a program to develop the Physical Agents Directive in 1990. The first draft of the Directive included guidance on noise, vibration, optical radiation, magnetic fields, and magnetic waves. This proved unpopular with all of the subdisciplines, and the program seemed quashed. However, in 1999, a new proposal for a series of Physical Agents Directives was accepted, with the intention of developing Directives for vibration and noise, and then using these as the basis for future Directives. A common position for the vibration Directive was reached in 2001, and the final version was published in the *Official Journal of the European Communities* on July 6, 2002. Throughout this process, the details of the Directive continually changed, including the limit values for whole-body vibration, in order to compromise between the demands of setting safe limits from a health perspective and pragmatism with respect to the practicalities of compliance with the Directive for some industrial sectors.

The Physical Agents (Vibration) Directive (2002) contains four sections comprising 16 articles. In addition, an Annex provides initial guidance on the methods for assessing hand-arm* and whole-body vibration in the workplace.

8.5.1 PHYSICAL AGENTS (VIBRATION) DIRECTIVE — SECTION 1: GENERAL PROVISIONS

Section 1 of the Directive contains three articles specifying the aim and scope of the document, definitions, and the exposure limit and action values.

The aim and scope of the Directive are to lay down the minimum requirements for protection of workers from health and safety risks due to mechanical vibration. Therefore, it does not apply to risks that might arise due to a leisure activity.

---

* The Physical Agents (Vibration) Directive uses the term *hand-arm vibration* rather than *hand-transmitted vibration*. The Directive's terminology is used in this section.

TABLE 8.3
Daily Exposure Limit Values and Exposure Action Values for Hand-Arm and Whole-Body Vibration, as Specified in the EU Physical Agents (Vibration) Directive (2002)

	Exposure Action Value	Exposure Limit Value
Hand-arm vibration	2.5 m/s^2 $A(8)$	5 m/s^2 $A(8)$
Whole-body vibration	0.5 m/s^2 $A(8)$	1.15 m/s^2 $A(8)$
	or	or
	9.1 m/s$^{1.75}$ VDV	21 m/s$^{1.75}$ VDV

Specifically, the Directive applies to activities in which exposure to health risks from vibration are likely.

It defines the terms *hand-arm vibration* and *whole-body vibration* in terms of risk. Therefore, according to the Directive, mechanical vibration transmitted to the whole body but not entailing risks to the health and safety of workers (such as found in the study of ride comfort in cars) should not be termed whole-body vibration. Disorders mentioned include vascular, skeletal, neurological, and muscular, which are related to hand-arm vibration, and low back morbidity and spine trauma for whole-body vibration.

At the core of the Directive are the exposure limit values and exposure action values (Table 8.3). The exposure of workers to vibration must be assessed or measured to enable comparison to be made with the limit and action values.

For hand-arm vibration, assessments are based on the daily exposure which is calculated as the root sum of the squares of the orthogonal frequency weighted vibration at the hands normalized to an 8-h reference period [$A(8)$]. Measurements should be made according to ISO 5349-1 (2001), using the W$_h$ frequency weighting. For tools that are held in two hands, the worse hand is used for the assessment (Figure 8.1).

For whole-body vibration, assessments are based on the daily exposure which is calculated as the most severe axis of vibration at the seat (or floor for a standing worker; see Figure 8.2). Assessments include axes multipliers as specified in ISO 2631 (1997). The quantity assessed should be either the frequency-weighted r.m.s. or, depending on the preference of the member state, the vibration dose value (VDV). Therefore, if making general assessments for a machine that is to be used across all of Europe, it is likely that both assessments will be required as some states will use r.m.s. and others VDV. Measurements should be made in accordance with ISO 2631. However, the Directive is clear about how to interpret the International Standard and, as a result, effectively removes many of the problems discussed in Subsection 6.4.2.

An alternative to making measurements of hand-arm or whole-body vibration is to estimate the exposure using the manufacturer's information regarding the vibration emission. This method should only be used when the manufacturer's method of obtaining the vibration emission values is representative of the situation

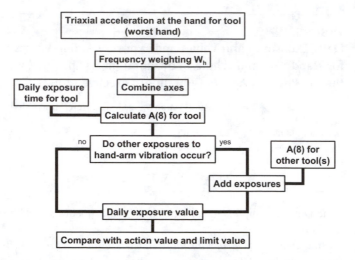

FIGURE 8.1 Method for assessing hand-arm vibration according to the EU Physical Agents (Vibration) Directive (2002).

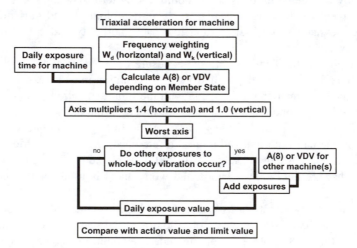

FIGURE 8.2 Method for assessing whole-body vibration according to the EU Physical Agents (Vibration) Directive (2002).

experienced by the operator at work; if exposures are close to (or exceed) the limit value, then an employer would be well advised also to make measurements of the vibration emission in the workplace.

8.5.2 PHYSICAL AGENTS (VIBRATION) DIRECTIVE — SECTION 2: OBLIGATION OF EMPLOYERS

Section 2 of the Directive contains four articles specifying the determination and assessment of risks, provisions to avoid or reduce exposure, worker information and training, and specifications for participation and consultation with affected workers.

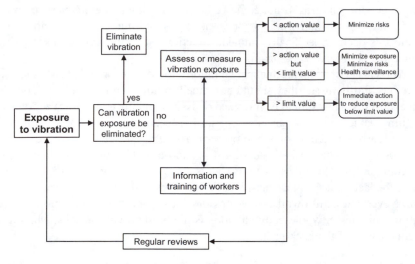

FIGURE 8.3 Action required for reducing or eliminating exposure for those at risk according to the EU Physical Agents (Vibration) Directive (2002).

Risk assessments should be carried out at regular intervals and appropriate records should be kept so that they can be consulted later. The assessments should not only consider the magnitudes of the vibration and the exposure action and limit values but also the nature of the exposures (e.g., exposure to mechanical shocks, intermittent vibration), issues associated with the health and safety of particularly sensitive workers, vibration as a confounding factor with other safety concerns, the existence of alternative methods or machinery, the effects of whole-body vibration outside of working hours, thermal effects and other information obtained from health surveillance, and the current state of knowledge. Although these factors are considered, there is no specific guidance on how to interpret the data. For example, if a worker participates in motor sport during leisure time, it is likely that the most severe vibration exposures occur during this leisure activity. Should the risk assessment identify this individual as being particularly at risk of injury or should the risk assessment identify that vibration at work is not the main contributor to the individual's total long-term exposure to vibration? It would not usually be appropriate for an employer to demand changes in leisure activity.

Risks from vibration exposures at work should be eliminated at source or reduced to a minimum (Figure 8.3). If the exposure action value is exceeded, measures must be implemented to minimize both exposure to vibration and the risks associated with the vibration. Risk reduction could be implemented as changes in the job design or machinery used. Alternatively, tools and machines could be modified so that less vibration is transmitted to the operator (e.g., by fitting vibration-reducing handles or seats). However, other ergonomic risks are highlighted in the Directive and must be considered for compliance. For example, the design and layout of workspaces should be optimized, and appropriate clothing must be provided if workers are exposed to cold and damp. Workers must also be trained to operate tools appropriately to minimize exposure. Ultimately, a reduction in the allowable duration of exposure could be enforced. Although employers are prohibited from exposing their workers above the exposure limit value, provision is made should this occur. Specifically, immediate action must be taken to ensure that the limit value is not exceeded again.

Workers at risk from mechanical vibration should be informed of risk assessment outcomes. Furthermore, they should be trained in how to minimize risks and vibration exposures. They should be informed on issues relating to exposure action values and limit values, know when they are entitled to health surveillance, learn how to identify early symptoms of vibration-related injury, and be encouraged to report these. Such self-diagnosis is possible for hand-arm vibration (e.g., incidence of blanching or loss of tactile sensitivity) but is more difficult for whole-body vibration due to the lack of a specific causal link between whole-body vibration and low back pain. Nevertheless, if a worker develops low back pain, then exposure to relatively high magnitudes of whole-body vibration is likely to provoke attacks, whatever the original cause of the back pain.

8.5.3 PHYSICAL AGENTS (VIBRATION) DIRECTIVE — SECTION 3: MISCELLANEOUS PROVISIONS

Section 3 of the Directive contains five articles specifying the implementation of health surveillance, when the Directive should be enforced, who it will affect, and mechanisms for making technical amendments to the document.

8.5.3.1 Health Surveillance

It is interesting that health surveillance is included in the "miscellaneous provisions" section rather than the "obligation of employers" section. Therefore, the responsibility falls on the EU member states to make provisions for health surveillance, although the Directive specifies that these provisions should be introduced in accordance with national procedures. Essentially, this means that in those states in which employers are responsible for health surveillance for other hazards at work, health surveillance for vibration will also be introduced as part of that same procedure.

The purpose of health surveillance is to identify the early indications of disorders associated with vibration so that those susceptible can be removed from the risk. The Directive states that:

> Where, as a result of health surveillance, a worker is found to have an identifiable disease or adverse health effect which is considered by a doctor or occupational health-care professional to be the result of exposure to mechanical vibration at work …

Therefore, doctors or health-care professionals have the duty to make the judgment as to whether vibration at work is the cause of the adverse health effect or not. Unfortunately, this is not always possible.

For hand-arm vibration, diseases are identifiable, and diagnostic testing and staging could be used to provide guidance as to how severe the disorder must be before the worker should be advised to change their job (see Chapter 4). For whole-body

TABLE 8.4
**Timetable for Implementation of the Physical Agents
(Vibration) Directive**

July 6, 2002	Directive published by EU
July 6, 2005	Implementation of Directive, excluding exposure limit values
July 6, 2007	Implementation of exposure limit values for new equipment only
July 6, 2010	Implementation of exposure limit values for all equipment
July 6, 2014	Implementation of exposure limit values for agriculture and forestry

vibration, there is less clarity. There is currently no clinically based technique available that will assist the assessor in determining the specific cause of any reported back pain.

If a worker is deemed to have a vibration-induced disorder, then the worker should be informed and advised on actions to take, including health surveillance, which should be continued after the exposure to vibration has stopped. The employer should also be informed so that improved risk assessments can be made and, if necessary, improved measures taken to reduce the risk.

8.5.3.2 Transitional Periods

The introduction of the Directive is to be phased. National laws enforcing the Directive must be in place by July 6, 2005 (Table 8.4). However, exposure limit values need not be implemented until July 6, 2007 for new equipment or until July 6, 2010 for old equipment. For the purposes of the Directive, new equipment is that which is given to workers after July 6, 2007. An extended transitional period applies for whole-body vibration in the agriculture and forestry sectors where the limit value need not be applied until July 6, 2014.

Although the transitional periods may appear attractive to some employers seeking to delay taking action, it must be stressed that action must be taken once the exposure action value is exceeded. Specifically, if other work equipment or methods reduce the risk by either reducing vibration emission or improving other ergonomic factors, then these should be used. However, the design of the Directive gives employers an incentive to purchase equipment prior to the July 6, 2007 cut-off or to delay purchasing equipment until after July 6, 2010, if there are doubts as to whether using the equipment would cause an exposure exceeding the limit value.

8.5.3.3 Derogations

For sea and air transport, member states may derogate compliance with the exposure-limit value for whole-body vibration, if it is not possible to comply. If derogations are granted by member states, then these must be temporary (less than 4 years), and workers should have closer health surveillance.

There are many individuals that are exposed to a widely varying vibration magnitude from day to day. In these cases it is possible to allow an $A(8)$ exposure to exceed the exposure limit value, provided that the $A(40)$ (i.e., an exposure averaged over a 40-h period) does not exceed the limit value. This derogation applies to both whole-body vibration and hand-arm vibration, but only when the vibration is usually below the action value. Some machines are used seasonally (e.g., some agricultural machines, snow grooming machines, some boats). The Directive, however, does not allow for averaging for longer periods than 40 h; therefore, they must comply with the $A(40)$, even if the worker's $A(\text{year})$ falls below the action value.

8.5.4 PHYSICAL AGENTS (VIBRATION) DIRECTIVE — SECTION 4: FINAL PROVISIONS

The final section of the Directive contains four articles concerned with the obligations on the member states for enforcement. It is required that member states shall report to the European Commission every 5 years on how the Directive has been implemented. Specifically, best practices should be identified such that these can be reported centrally and, if necessary, amendments made to the document.

8.5.5 REVIEW OF THE PHYSICAL AGENTS (VIBRATION) DIRECTIVE

The Directive's aim is to improve the health and safety of workers across Europe. As it is accepted that exposure to mechanical vibration can be hazardous, measures to reduce the vibration should reduce the incidence of vibration-related diseases. As such, it must be welcomed. Nevertheless, there are considerable implications across many industries, ultimately affecting operational costs and, in some cases, business viability. Inevitably, the details of the legal framework will be confirmed through case law in each country.

8.5.5.1 Financial Costs

The Physical Agents (Vibration) Directive will incur potentially high costs on European industry. Costs will come from carrying out risk assessments, modification or replacement of machinery, health surveillance, and training of workers. Further indirect costs may come if the increased profile of whole-body vibration as a risk for back pain stimulates an increase in claims for damages. It is estimated that the largest costs for hand-arm vibration will come from health surveillance and reducing exposures; for whole-body vibration, most of the cost will come from measures to reduce exposure (Coles, 2002).

The costs associated with compliance must be balanced with the benefits. If the Directive is a success, then there will, in the future, be a lower incidence of vibration-related illnesses. Therefore, the burden on the medical sector will be reduced, and there will be fewer claims for compensation falling on the business and/or insurance sector. The improvement in the quality of life of those that have been, unknowingly, protected from the potential distress from vibration-related disorders is impossible to quantify financially.

Although replacing equipment can be costly, the new equipment must be manufactured and therefore suppliers will benefit from the Directive, possibly creating new jobs within the affected countries. Should this occur, then the overall financial impact on society as a whole will be reduced. A similar argument can be used for all other costs incurred to enable compliance with the directive (e.g., the need for extra health professionals, vibration consultants, ergonomists, and individuals to carry out health surveillance, etc.).

8.5.5.2 Improvements in Product Design

One of the implications of the Physical Agents (Vibration) Directive is that the vibration emission values of machinery will become increasingly important to machine purchasers. Therefore, manufacturers have an incentive to provide equipment that produces the lowest vibration magnitudes. As vibration-reducing technologies improve, fewer high-vibration machines should be on the market. Therefore, all users will benefit from the reduced-vibration devices, irrespective of whether or not they use them for long periods of time and are considered to be at risk.

The Directive refers to "auxiliary equipment" that can be fitted to machines to reduce the risk of injuries caused by vibration. Due to the increased market for such equipment, a larger budget for research could be invested by manufacturers and improvements in technology could result, be it for suspension seats, vibration-reducing handles, or antivibration gloves.

8.5.5.3 Are the Methods Correct?

For hand-arm vibration, the methods specified in the Directive correspond with the generally accepted best practice for vibration assessment, although there is some doubt whether W_h is a better predictor of risk of developing vibration-induced white finger than to use unweighted acceleration (e.g., Griffin et al., 2003). The specification of using the root sum of squares of the acceleration in all axes is a step forwards, especially when compared to some previous guidance that specified that the worst axis should be considered (e.g., HSE guidelines, 1994).

It is unfortunate that the methods specified for whole-body vibration dictate that the worst axis of vibration is the used for the risk assessment value to be considered. This is a failing of the Directive; one could argue that all axes need to be measured, anyway, to determine which axis dominates. If one axis proved to be dominant, the others will have little influence on the overall vibration magnitude. However, if the three directions of vibration are similar, then the root sum of squares for the r.m.s. vibration exposure could, theoretically, be 73% higher than that indicated in the worst axis only. This also amounts to an inconsistency, as the hand-arm vibration methods sum the axes.

Member states have the option to use r.m.s. or VDV for the exposure action values and limit values. It is likely that most states will opt for r.m.s. methods, although some will implement VDV, at least for the action value. Although VDV is generally considered to be a better indicator of risk (as it emphasizes the shocks in the vibration), it is more difficult to measure and interpret. Furthermore, the Machinery Safety Directive only requires manufacturers to provide emission values in terms of r.m.s.; therefore, it is difficult for purchasers to select appropriate low-emission machinery when measured as VDV. The r.m.s. or VDV option is likely to result in a lack of harmonization across Europe, so that some whole-body vibration exposures will cross the exposure action or limit value threshold depending on the member state.

8.5.5.4 Are the Exposure Limit Values and Exposure Action Values Correct?

The ultimate goal of the Directive is to minimize and, where possible, avoid damage to workers' health. Unfortunately, interindividual variation in susceptibility and imperfect vibration monitoring mean that the ultimate target of a perfect predictor of injury is impossible to achieve. Therefore, it might be tempting to err on the side of caution when setting limits. However, if the limits are set too low, then some tasks with a low risk might be prohibited, and hence industry would bear financial penalties with little improvement in health. Conversely, if the limits are set too high, then action that could protect health might not be taken by employers.

A dose–effect relationship has been standardized for hand-arm vibration exposure, and so percentages of persons with adversely affected health can be estimated from measures of vibration. For whole-body vibration, no such relationship is established and so setting limits is more difficult. One reason for this is that vibration is just one of a range of factors that might lead to back pain. In practice, most processes should be able to comply with the whole-body vibration limit values if modern methods and machines are used. The limit value should effectively ensure that the worst of the older machinery is phased out.

The limit and action values for hand-arm vibration in the Physical Agents (Vibration) Directive can be used with the models for blanching in ISO 5349 (2001). This model predicts that 10% of those exposed at the limit value would experience finger blanching after 6 years. Likewise, 10% of those exposed at the action value would experience finger blanching after 12 years. There are many tools that would exceed the limit value if used for a full working day. However, most hand tools are not used continuously and strict time limits might be required, especially for pneumatic impact tools where it is not foreseeable that they will comply with the Directive for a full 8-h exposure.

To concentrate on the exposure action values and limit values somewhat diverts focus from the general requirements of the Directive. Vibration exposures must be reduced for all of those at risk, irrespective of the vibration magnitude. Reducing risk should result in a lower incidence of vibration related disorders in the long term.

8.6 THE PHYSICAL AGENTS (NOISE) DIRECTIVE: DIRECTIVE 2003/10/EC OF THE EUROPEAN PARLIAMENT AND OF THE COUNCIL ON THE MINIMUM HEALTH AND SAFETY REQUIREMENTS REGARDING THE EXPOSURE OF WORKERS TO THE RISKS ARISING FROM PHYSICAL AGENTS (NOISE)

The Physical Agents (Noise) Directive is analogous to the vibration Directive. It also has similarities to the previous European noise Directive (European Commission, 1986) that was implemented across member states (e.g., as the Noise at Work Regulations in the U.K., HMSO, 1989).

TABLE 8.5
Daily Exposure Limit Values and Action Values for
Noise as Specified in the EU Physical Agents (Noise)
Directive (2003)

	Lower Exposure Action Value	Upper Exposure Action Value	Exposure Limit Value
8-h equivalent (L_{Aeq})	80 dB(A)	85 dB(A)	87 dB(A)
Peak pressure	112 Pa [135 dB(C)]	140 Pa [137 dB(C)]	200 Pa [140 dB(C)]

The main changes from the previous legislation are that the equivalent action levels and limit values have been reduced (Table 8.5). At the lower exposure action value, hearing protection must be made available and training must be provided. At the upper exposure action value, suitable hearing protection must be worn, workers are entitled to audiometric testing, a program of control measures must be implemented and workplaces where workers are likely to exceed the value should be marked with appropriate signs. The exposure limit value must not be exceeded. This means that if a worker is provided with hearing protection as the noise exceeds 85 dB(A), the protection must reduce the noise at the ear to a maximum of 87 dB(A). Where noise levels change from day to day, a weekly average is allowable.

The noise Directive will come into force on February 15, 2006. However, the music and entertainment sectors do not need to comply until 2008, and personnel on sea vessels do not need to comply with the limit value until 2011.

Further consideration of the Physical Agents (Noise) Directive is beyond the scope of this book.

8.7 CHAPTER SUMMARY

European Directives must be implemented in the law of the member states of the European Union and European Economic Area. One of their purposes is to ensure a common degree of health and safety across all states, such that industries in one part of Europe cannot cut costs by compromising health and safety.

The Machinery Safety Directive gives minimum requirements for many aspects of the safety of powered devices. Machines must be designed such that the risks of vibration related injury are minimized. However, if the vibration magnitudes exceed thresholds for hand-transmitted and whole-body vibration, then these magnitudes must be stated in the instruction book.

The Physical Agents (Vibration) Directive will place legal limits on the vibration exposures of workers across all European member states. These limits are expected to affect several million workers. All employers will be required to assess and, if necessary, measure the vibration exposures of their employees to identify whether hand-arm vibration or whole-body vibration exposures exceed the exposure limit and exposure action values. Many operations will require modification, and it is likely that many machines will need to be replaced with new low-vibration models.

References

Agarwal, A., Pathak, A., and Gaur, A. (2000). Acupressure wristbands do not prevent post-operative nausea and vomiting after urological endoscopic surgery. *Canadian Journal of Anaesthesia,* 47(4), 319–324.

Åreskoug, A., Hellström, P.-A., Lindén, B., Kähäri, K., Zachau, G., Olsson, J., Häll, A., Sjösten, P., and Forsman M. (2000). Auto-balancing on angle grinders. Proceedings of 8th International Conference on Hand-Arm Vibration, June 9–12, 1998, Umeå, Sweden.

Atkinson, S., Robb, M.J.M., and Mansfield, N.J. (2002). Long term vibration dose for truck drivers: preliminary results and methodological challenges. Proceedings of 37th U.K. Conference on Human Response to Vibration, held at the Department of Human Sciences, Loughborough University, September 18–20, 2002.

Bagshaw, M. and Stott, J.R.R. (1985). The desensitisation of chronically motion sick aircrew in the Royal Air Force. *Aviation, Space and Environmental Medicine,* 56, 1144–1151.

Bendat, J.S. and Piersol, A.G. (1986). *Random Data: Analysis and Measurement Procedures.* New York: John Wiley & Sons.

Benson, A.J. (1984). Motion sickness. In *Vertigo*, Dix, M.R. and Hood, J.D., Eds. New York: John Wiley & Sons.

Benson, A.J. and Dilnot, S. (1981). Perception of whole-body linear oscillation. Proceedings of U.K. Informal Group on Human Response to Vibration, Heriot-Watt University, Edinburgh, September 9–11, 1981.

Boshuizen, H.C., Hulshof, C.T., and Bongers, P.M. (1990). Long-term sick leave and disability pensioning due to back disorders of tractor drivers exposed to whole-body vibration. *International Archives of Occupational and Environmental Health,* 62(2), 117–122.

Bovenzi, M. (1998). Exposure-response relationship in the hand-arm vibration syndrome: an overview of current epidemiology research. *International Archives of Occupational and Environmental Health,* 71(8), 509–519.

Bovenzi, M. (2002). Finger systolic blood pressure indices for the diagnosis of vibration-induced white finger. *International Archives of Occupational and Environmental Health,* 75(1–2), 20–28.

Bovenzi, M. and Betta, A. (1994). Low-back disorders in agricultural tractor drivers exposed to whole-body vibration and postural stress. *Applied Ergonomics,* 25(4), 231–241.

Bovenzi, M. and Hulshof, C.T.J. (1998). An updated review of epidemiologic studies on the relationship between exposure to whole-body vibration and low back pain. *Journal of Sound and Vibration,* 215(4), 595–612.

Bovenzi, M. and Hulshof, C.T.J. (1999). An updated review of epidemiologic studies on the relationship between exposure to whole-body vibration and low back pain (1986–1997). *International Archives of Occupational and Environmental Health,* 72(6), 351–65.

Bradford-Hill, A. (1966). The environment and disease: association or causation? Proceedings of the Royal Society of Medicine, 58, 295.

Brammer, A.J. (1986). Dose-response relationships for hand-transmitted vibration. *Scandinavian Journal of Work and Environmental Health,* 12, 284–288.

Brammer, A.J., Taylor, W., and Lundborg, G. (1987). Sensorineural stages of the hand-arm vibration syndrome. *Scandinavian Journal of Work and Environmental Health,* 13, 279–283.

Brandt, T. (1999). *Vertigo: Its Multisensory Syndromes.* (2nd ed.). London: Springer.

Brandt, T., Wenzel, D., and Dichgans, J. (1976). Die Entwicklung der visuellen Stabilisation des aufrechten Standes beim Kind: Ein Reifezeichen in der Kinderneurologie (Visual stabilization of free stance in infants: a sign of maturity). *Archiv fur Psychiatrie und Nervenkrankheiten,* 223(1), 1–13.

Bridger, R.S., Groom, M.R., Jones, H., Pethybridge, R.J., and Pullinger, N.C. (2002). Back pain in Royal Navy helicopter pilots, part 2: some methodological considerations. *Contemporary Ergonomics 2002,* P.T. McCabe, Ed. London: Taylor & Francis.

Bridger, R.S., Groom, M.R., Pethybridge, R.J., Pullinger, N.C., and Paddan, G.S. (2002). Back pain in Royal Navy helicopter pilots, part 1: general findings. *Contemporary Ergonomics 2002,* P.T. McCabe, Ed. London: Taylor & Francis.

British Standards Institution (1974). Guide to the evaluation of human exposure to whole-body vibration. BSI DD 32. London: British Standards Institution.

British Standards Institution (1975). Guide to the evaluation of exposure of the human hand-arm system to vibration. BSI DD 43. London: British Standards Institution.

British Standards Institution (1984). Evaluation of human exposure to vibration in buildings (1 Hz to 80 Hz). BS 6472. London: British Standards Institution.

British Standards Institution (1985). Evaluation of the response of occupants of fixed structures, especially buildings and offshore structures, to low frequency horizontal motion (0.063 to 1 Hz). BS 6611. London: British Standards Institution.

British Standards Institution (1987). Measurement and evaluation of human exposure to vibration transmitted to the hand. BS 6842. London: British Standards Institution.

British Standards Institution (1987). Measurement and evaluation of human exposure to whole-body mechanical vibration and repeated shock. BS 6841. London: British Standards Institution.

British Standards Institution (1993). Guide to the evaluation of exposure of the human hand-arm system to vibration. DD ENV 25349. London: British Standards Institution.

British Standards Institution (1993). Hand-held portable power tools: measurement of vibrations at the hands — part 1: general. BS 28662-1. London: British Standards Institution.

British Standards Institution (1994). Mechanical vibration: laboratory method for evaluating vehicle seat vibration — part 1: basic requirements. BS 303261-1. London: British Standards Institution.

British Standards Institution (1995). Hand-held portable power tools: measurement of vibrations at the handle — part 4: grinding machines. BS 8662-4. London: British Standards Institution.

British Standards Institution (1995). Hand-held portable power tools: measurement of vibrations at the handle — part 6: impact drills. BS 8662-6. London: British Standards Institution.

British Standards Institution (1995). Hand-held portable power tools: measurement of vibrations at the hands — part 2: chipping hammers and riveting hammers. BS 28662-2. London: British Standards Institution.

British Standards Institution (1996). Hand-arm vibration: guidelines for vibration hazards reduction — part 2: management measures at the workplace. PD 6585-2. London: British Standards Institution.

British Standards Institution (1996). Hand-arm vibration: laboratory measurement of vibration at the grip surface of hand-guided machinery — general. BS 1033. London: British Standards Institution.

British Standards Institution (1997). Mechanical vibration and shock: hand-arm vibration — method for the measurement and evaluation of the vibration transmissibility of gloves at the palm of the hand. BS10819. London: British Standards Institution.

British Standards Institution (1997). Mechanical vibration: testing of mobile machinery in order to determine the whole-body vibration emission value — general. BS 1032. London: British Standards Institution.

British Standards Institution (1999). Mechanical vibration and shock: hand-arm vibration — method for measuring the vibration transmissibility of resilient materials when loaded by the hand-arm system. BS 13753. London: British Standards Institution.

British Standards Institution (2000). Earth moving machinery: laboratory evaluation of operator seat vibration. BS 7096. London: British Standards Institution.

British Standards Institution (2000). Earth moving machinery: laboratory evaluation of operator seat vibration. BS 8912-17. London: British Standards Institution.

British Standards Institution (2001). Mechanical vibration: laboratory method for evaluating vehicle seat vibration — part 2: application to railway vehicles. BS 10326-2. London: British Standards Institution.

British Standards Institution (2001). Mechanical vibration: measurement and analysis of whole-body vibration to which passengers and crew are exposed in railway vehicles. BS 10056. London: British Standards Institution.

British Standards Institution (2001). Mechanical vibration: measurement and evaluation of human exposure to hand transmitted vibration — part 1: general guidelines. BS 5349-1. London: British Standards Institution.

British Standards Institution (2002). Agricultural wheeled tractors and field machinery: measurement of whole-body vibration of the operator. BS 5008. London: British Standards Institution.

British Standards Institution (2002). Mechanical vibration: industrial trucks — laboratory evaluation and specification of operator seat vibration. BS 13490. London: British Standards Institution.

British Standards Institution (2002). Mechanical vibration: measurement and evaluation of human exposure to hand transmitted vibration — part 2: practical guidance for measurement at the workplace. BS 5349-2. London: British Standards Institution.

British Standards Institution (2002). Safety of industrial trucks: test methods for measuring vibration. BS 13059. London: British Standards Institution.

British Standards Institution (2003). Mechanical vibration: testing of mobile machinery in order to determine the vibration emission value. BS 1032. London: British Standards Institution.

Bruce, D.G., Golding, J.F., Hockenhull, N., and Pethybridge, R.J. (1990). Acupressure and motion sickness. *Aviation, Space and Environmental Medicine,* 61(4), 361–365.

Carr C. (2001). Overview of the motion sickness desensitisation programme at the Centre of Human Sciences. Proceedings of the 36th U.K. Group Conference on Human Response to Vibration, held at Centre for Human Science, QinetiQ, Farnborough, U.K., September 12–14, 2001.

Castelo-Branco, N.A. (1999). The clinical stages of vibroacoustic disease. *Aviation, Space and Environmental Medicine,* 70(3), A32–A39.

Castelo-Branco, N.A. and Rodriguez, E. (1999). The vibroacoustic disease: an emerging pathology. *Aviation, Space and Environmental Medicine,* 70(3), A1–A6.

The Chartered Society of Physiotherapy (2001). *Take the Pain Out of Driving.* London: The Chartered Society of Physiotherapy.

Clemes, S.A. and Howarth, P.A. (2002). Evidence of habituation to virtual simulator sickness when volunteers are tested at weekly intervals. Proceedings of the 37th U.K. Conference on Human Response to Vibration, held at the Department of Human Sciences, Loughborough University, September 18–20, 2002.

Coles, B. (2002). Regulatory impact assessment of the Physical Agents (Vibration) Directive. Health and Safety Executive.

Corbridge, C. (1981). Effect of subject weight on suspension seat vibration transmissibility: suspension in and out. Proceedings of U.K. Informal Group on Human Response to Vibration, Heriot-Watt University, Edinburgh, September 9–11, 1981.

Corbridge, C., Griffin, M.J., and Harborough, P.R. (1989). Seat dynamics and passenger comfort. Proceedings of the Institute of Mechanical Engineers 203, 57–64.

Coren, S., Ward, L.M., and Enns, J.T. (1999). *Sensation and Perception*. (5th ed.). New York: John Wiley & Sons.

Council of the European Communities (1983). Council Directive 83/189/EEC of 28 March 1983 laying down a procedure for the provision of information in the field of technical standards and regulations. *Official Journal of the European Communities*, L109.

Council of the European Communities (1989). Council Directive 89/392/EEC on the approximation of the laws of the member states relating to machinery. *Official Journal of the European Communities*, L183.

Council of the European Communities (1998). Council Directive 98/37/EC on the approximation of the laws of the member states relating to machinery. *Official Journal of the European Communities*, L207.

Crampton, G.H. (1990). Motion and Space Sickness. Boca Raton, Florida: CRC Press.

Crowley, J.S. (1987). Simulator sickness: a problem for army aviation. *Aviation, Space and Environmental Medicine*, 58(4), 355–357.

Cullmann, A. and Wölfel, H.P. (2001). Design of an active vibration dummy of sitting man. *Clinical Biomechanics*, 16 Suppl 1:S64–S72.

Darwin, E. (1796). *Zoonomia, or The Laws of Organic Life*.

Drerup, B., Granitzka, M., Assheuer, J., and Zerlett, G. (1999). Assessment of disc injury in subjects exposed to long-term whole-body vibration. *European Spine Journal*, 8(6), 458–467.

Ebe, K. and Griffin, M.J. (2000a). Qualitative models of seat discomfort including static and dynamic factors. *Ergonomics*, 43(6), 771–790.

Ebe, K. and Griffin, M.J. (2000b). Quantitative prediction of overall seat discomfort. *Ergonomics*, 43(6), 791–806.

Ebe, K. and Griffin, M.J. (2001). Factors affecting static seat cushion comfort. *Ergonomics*, 44(10), 901–921.

European Commission (1986). Council Directive 86/188/EEC of 12 May 1986 on the protection of workers from the risks related to exposure to noise at work. *Official Journal of the European Communities*, L137.

European Commission (1990). Council Directive 90/269/EEC of 29 May 1990 on the minimum health and safety requirements for the manual handling of loads where there is a risk particularly of back injury to workers (fourth individual Directive within the meaning of Article 16 (1) of Directive 89/391/EEC). *Official Journal of the European Communities*, L156.

European Commission (2002). Directive 2002/44/EC of the European Parliament and of the Council of 25 June 2002 on the minimum health and safety requirements regarding the exposure of workers to the risks arising from physical agents (vibration) (sixteenth individual Directive within the meaning of Article 16(1) of Directive 89/391/EEC). *Official Journal of the European Communities*, L177.

European Commission (2003). Directive 2003/10/EC of the European Parliament and of the Council of 6 February 2003 on the minimum health and safety requirements regarding the exposure of workers to the risks arising from physical agents (noise) (seventeenth individual Directive within the meaning of Article 16(1) of Directive 89/391/EEC). *Official Journal of the European Communities*, L42.

European Committee for Standardization (1992). Hand-held portable power tools: measurement of vibrations at the hands — part 1: general. EN 28662-1. Brussels: European Committee for Standardization.

European Committee for Standardization (1994). Hand-held portable power tools: measurement of vibrations at the hands — part 2: chipping hammers and riveting hammers. EN 28662-2. Brussels: European Committee for Standardization.

European Committee for Standardization (1994). Mechanical vibration: laboratory method for evaluating vehicle seat vibration — part 1: basic requirements. EN 30326-1. Brussels: European Committee for Standardization.

European Committee for Standardization (1995). Hand-arm vibration: laboratory measurement of vibration at the grip surface of hand-guided machinery — general. EN 1033. Brussels: European Committee for Standardization.

European Committee for Standardization (1995). Hand-held portable power tools: measurement of vibrations at the handle — part 6: impact drills. EN 8662-6. Brussels: European Committee for Standardization.

European Committee for Standardization (1995). Hand-held portable power tools: measurement of vibrations at the hands — Part 4: grinding machines. EN 8662-4. Brussels: European Committee for Standardization.

European Committee for Standardization (1996). Mechanical vibration and shock: hand-arm vibration — method for the measurement and evaluation of the vibration transmissibility of gloves at the palm of the hand. EN 10819. Brussels: European Committee for Standardization.

European Committee for Standardization (1996). Mechanical vibration: testing of mobile machinery in order to determine the whole-body vibration emission value — general. EN 1032. Brussels: European Committee for Standardization.

European Committee for Standardization (1998). Mechanical vibration and shock: guidance on safety aspects of tests and experiments with people — exposure to whole-body mechanical vibration and repeated shock. EN 13090-1. Brussels: European Committee for Standardization.

European Committee for Standardization (1998). Mechanical vibration and shock: hand-arm vibration — method for measuring the vibration transmissibility of resilient materials when loaded by the hand-arm system. EN 13753. Brussels: European Committee for Standardization.

European Committee for Standardization (2000). Earth moving machinery: laboratory evaluation of operator seat vibration. EN 7096. Brussels: European Committee for Standardization.

European Committee for Standardization (2001). Mechanical vibration: industrial trucks — laboratory evaluation and specification of operator seat vibration. EN 13490. Brussels: European Committee for Standardization.

European Committee for Standardization (2001). Mechanical vibration: measurement and evaluation of human exposure to hand transmitted vibration — part 1: general guidelines. EN 5349-1. Brussels: European Committee for Standardization.

European Committee for Standardization (2002). Mechanical vibration: measurement and evaluation of human exposure to hand transmitted vibration — part 2: practical guidance for measurement at the workplace. EN 5349-2. Brussels: European Committee for Standardization.

European Committee for Standardization (2002). Safety of industrial trucks: test methods for measuring vibration. EN 13059. Brussels: European Committee for Standardization.

European Committee for Standardization (2003). Mechanical vibration: testing of mobile machinery in order to determine the vibration emission value. EN 1032. Brussels: European Committee for Standardization.

Fahy, F.J. and Walker, J.G. (1998). *Fundamentals of Noise and Vibration*. London: E&FN Spon.

Fairley, T.E. (1983). The effect of vibration characteristics on seat transmissibility. Proceedings of the U.K. Informal Group on Human Response to Vibration, NIAE/NCAE, Silsoe, Beds, September 14–16, 1983.

Fairley, T.E. and Griffin, M.J. (1989). The apparent mass of the seated human body: vertical vibration. *Journal of Biomechanics*, 22(2), 81–94.

Fairley, T.E. and Griffin, M.J. (1990). The apparent mass of the seated human body in the fore-and-aft and lateral directions. *Journal of Sound and Vibration*, 139(2), 299–306.

Farkkila, M.A., Pyykko, I., Starck, J.P., and Korhonen, O.S. (1982). Hand grip force and muscle fatigue in the etiology of the vibration syndrome. In *Vibration Effects on the Hand and Arm in Industry*, Brammer, A.J. and Taylor, W., Eds. New York: John Wiley & Sons.

Förstberg, J. and Kufver, B.K. (2001). Ride comfort and motion sickness in tilting trains. Proceedings of the 36th U.K. Group Conference on Human Response to Vibration, held at Centre for Human Sciences, QinetiQ, Farnborough, U.K., September 12–14, 2001.

Futatsuka, M. and Fukuda, Y. (2000). A follow up study on the consequences of VWF patients in the workers using chain saws. Proceedings of 8th International Conference on Hand-Arm Vibration, June 9–12, 1998, Umeå, Sweden.

Gemne, G., Pyykko, I., Taylor, W., and Pelmear, P. (1987). The Stockholm Workshop scale for the classification of cold-induced Raynaud's phenomenon in the hand-arm vibration syndrome (revision of the Taylor-Pelmear scale). *Scandinavian Journal of Work and Environmental Health*, 13(4), 275–278.

Golding, J.F., Finch, M.I., and Stott, J.R. (1997). Frequency effect of 0.35-1.0 Hz horizontal translational oscillation on motion sickness and the somatogravic illusion. *Aviation, Space and Environmental Medicine*, 68(5), 396–402.

Golding, J.F. and Markey, H.M. (1996). Effect of frequency of horizontal linear oscillation on motion sickness and somatogravic illusion. *Aviation, Space and Environmental Medicine*, 67(2), 121–126.

Golding, J.F., Markey, H.M., and Stott, J.R. (1995). The effects of motion direction, body axis and posture on motion sickness induced by low frequency linear oscillation. *Aviation, Space and Environmental Medicine*, 66(11), 1046–1051.

Golding, J.F., Mueller, A.G., and Gresty, M.A. (2001). A motion sickness maximum around the 0.2 Hz frequency range of horizontal translational oscillation. *Aviation, Space and Environmental Medicine*, 72(3), 188–192.

Golding, J.F. and Stott, J.R. (1995). Effect of sickness severity on habituation to repeated motion challenges in aircrew referred for airsickness treatment. *Aviation, Space and Environmental Medicine*, 66(7), 625–630.

Griffin, M.J. (1990). *Handbook of Human Vibration*. London: Academic Press.

Griffin, M.J. (1998a). A comparison of standardized methods for predicting the hazards of whole-body vibration and repeated shocks. *Journal of Sound and Vibration*, 215(4), 883–914.

Griffin, M.J. (1998b). Evaluating the effectiveness of gloves in reducing the hazards of hand-transmitted vibration. *Occupational and Environmental Medicine*, 55(5), 340–348.

Griffin, M.J. (1998c). Fundamentals of human responses to vibration. In *Fundamentals of Noise and Vibration*, Fahy, F.J. and Walker, J.G., Eds. London: E&FN Spon.

Griffin, M.J. (2001). The validation of biodynamic models. *Clinical Biomechanics*, 16(1), S81–S92.

Griffin, M.J., Bovenzi, M., and Nelson, C.M. (2003). Dose-response patterns for vibration-induced white finger. *Occupational and Environmental Medicine*, 60(1), 16–26.

Griffin, M.J. and Mills K.L. (2002). Effect of frequency and direction of horizontal oscillation on motion sickness. *Aviation, Space and Environmental Medicine*, 73(7), 640–646.

Griffin, M.J. and Whitham, E.M. (1980). Time dependency of whole-body vibration discomfort. *Journal of the Acoustical Society of America*, 68(5), 1522–1523.

Griffin, M.J., Whitham, E.M., and Parsons, K.C. (1982). Vibration and comfort, I: translational seat vibration. *Ergonomics*, 25(7), 603–630.

Gyi, D.E. and Porter, J.M. (1999). Interface pressure and the prediction of car seat discomfort. *Applied Ergonomics*, 30(2), 99–107.

Hammond, J.K. (1998). Fundamentals of signal processing. In *Fundamentals of Noise and Vibration*, Fahy, F.J. and Walker, J.G., Eds. London: E&FN Spon.

Harada N. (2002). Cold-stress tests involving finger skin temperature measurement for evaluation of vascular disorders in hand-arm vibration syndrome: review of the literature. *International Archives of Occupational and Environmental Health*, 75(1–2), 14–19.

Harada, N. and Griffin, M.J. (1991). Factors influencing vibration sense thresholds used to assess occupational exposures to hand transmitted vibration. *British Journal of Industrial Medicine*, 48(3), 185–192.

Harmon, D., Ryan, M., Kelly, A., and Bowen, M. (2000). Acupressure and prevention of nausea and vomiting during and after spinal anaesthesia for caesarean section. *British Journal of Anaesthesia*, 84(4), 463–467.

Harris, C.M. and Piersol, A.G. (2002). *Harris' Shock and Vibration Handbook*. (5th ed.). New York: McGraw-Hill.

Health and Safety Executive (1994). *Hand-Arm Vibration*, HSG88 Sudbury, U.K.: HSE Books.

Hedlund, U. (1989). Raynaud's phenomenon of fingers and toes of miners exposed to local and whole-body vibration and cold. *International Archives of Occupational and Environmental Health*, 61(7), 457–461.

Her Majesty's Stationery Office (1989). The noise at work regulations. Statutory Instrument No. 1790. London: HMSO.

Her Majesty's Stationery Office (1992). The manual handling operations regulations. Statutory Instrument No. 2793. London: HMSO.

Hewitt, S. (2002). Round robin testing of anti-vibration gloves towards a revision of ISO10819:1996. Proceedings of 37th United Kingdom Conference on Human Response to Vibration, held at the Department of Human Sciences, Loughborough University, September 18–20, 2002.

Hill, K.J. and Howarth, P.A. (2000). Habituation to the side effects of immersion in a virtual environment. *Displays*, 21(1), 25–30.

Holmes, S.R., King, S., Stott, J.R.R., and Clemes, S. (2002). Facial skin pallor increases during motion sickness. *Journal of Psychophysiology*, 16(3), 150–157.

Holmlund, P. and Lundström, R. (1998). Mechanical impedance of the human body in the horizontal direction. *Journal of Sound and Vibration*, 215(4), 801–812.

Howarth, H.V.C. and Griffin, M.J. (1988). The frequency dependence of subjective reaction to vertical and horizontal whole-body vibration at low magnitudes. *Journal of the Acoustical Society of America*, 83(4), 1406–1413.

Howarth, P.A. and Hill, K.J. (1999). The maintenance of habituation to virtual simulation sickness. Proceedings of 8th International Conference on Human Computer Interaction. Munich, Germany. August 22–26, 1999.

Hulshof, C. and van Zanten, B.V. (1987). Whole-body vibration and low-back pain: a review of epidemiologic studies. *International Archives of Occupational and Environmental Health,* 59, 205–220.

Huston, D.R., Zhao, X.D., and Johnson, C.C. (2000). Whole-body shock and vibration: frequency and amplitude dependence of comfort. *Journal of Sound and Vibration,* 230(4), 964–970.

International Organization for Standardization (1979). Agricultural wheeled tractors and field machinery: measurement of whole-body vibration of the operator. ISO 5008. Geneva: International Organization for Standardization.

International Organization for Standardization (1984). Guidelines for the evaluation of the response of occupants of fixed structures, especially buildings and off-shore structures, to low-frequency horizontal motion (0.063 to 1 Hz). ISO 6897. Geneva: International Organization for Standardization.

International Organization for Standardization (1985). Evaluation of human exposure to whole-body vibration: part 1 — general requirements. ISO 2631/1. Geneva: International Organization for Standardization.

International Organization for Standardization (1985). Evaluation of human exposure to whole-body vibration: part 3 — evaluation of exposure to whole-body z-axis vertical vibration in the frequency range 0.1 to 0.63 Hz. ISO 2631-3. Geneva: International Organization for Standardization.

International Organization for Standardization (1986). Forestry machinery: chain saws — measurement of hand-transmitted vibration. ISO 7505. Geneva: International Organization for Standardization.

International Organization for Standardization (1986). Mechanical vibration: guidelines for the measurement and the assessment of human exposure to hand-transmitted vibration. ISO 5349. Geneva: International Organization for Standardization.

International Organization for Standardization (1988). Hand-held portable power tools: measurement of vibrations at the handle — part 1: general. ISO 8662-1. Geneva: International Organization for Standardization.

International Organization for Standardization (1989). Evaluation of human exposure to whole-body vibration: part 2 — continuous and shock induced vibration in buildings (1 to 80 Hz). ISO 2631-2. Geneva: International Organization for Standardization.

International Organization for Standardization (1989). Forestry machinery: portable brushsaws — measurement of hand-transmitted vibration. ISO 7916. Geneva: International Organization for Standardization.

International Organization for Standardization (1990). Human response to vibration: measuring instrumentation. ISO 8041. Geneva: International Organization for Standardization.

International Organization for Standardization (1992). Hand-held portable power tools: measurement of vibrations at the handle — part 2: chipping hammers and riveting hammers. ISO 8662-2. Geneva: International Organization for Standardization.

International Organization for Standardization (1992). Hand-held portable power tools: measurement of vibrations at the handle — part 3: rock drills and rotary hammers. ISO 8662-3. Geneva: International Organization for Standardization.

International Organization for Standardization (1992). Hand-held portable power tools: measurement of vibrations at the handle — part 5: pavement breakers and hammers for construction work. ISO 8662-5. Geneva: International Organization for Standardization.

International Organization for Standardization (1992). Mechanical Vibration: laboratory method for evaluating vehicle seat vibration — part 1: basic requirements. ISO 10326. Geneva: International Organization for Standardization.

International Organization for Standardization (1994). Earth moving machinery: laboratory evaluation of operator seat vibration. ISO 7096. Geneva: International Organization for Standardization.

International Organization for Standardization (1994). Hand-held portable power tools: measurement of vibrations at the handle — part 4: grinders. ISO 8662-4. Geneva: International Organization for Standardization.

International Organization for Standardization (1994). Hand-held portable power tools: measurement of vibrations at the handle — part 6: impact drills. ISO 8662-6. Geneva: International Organization for Standardization.

International Organization for Standardization (1996). Hand-held portable power tools: measurement of vibrations at the handle — part 9: rammers. ISO 8662-9. Geneva: International Organization for Standardization.

International Organization for Standardization (1996). Hand-held portable power tools: measurement of vibrations at the handle — part 14: stone-working tools and needle scalers. ISO 8662-14. Geneva: International Organization for Standardization.

International Organization for Standardization (1996). Mechanical vibration and shock: hand-arm vibration — method for the measurement and evaluation of the vibration transmissibility of gloves at the palm of the hand. ISO 10819. Geneva: International Organization for Standardization.

International Organization for Standardization (1997). Hand-held portable power tools: measurement of vibrations at the handle — part 12: saws and files with reciprocating action and saws with oscillating or rotating action. ISO 8662-12. Geneva: International Organization for Standardization.

International Organization for Standardization (1997). Hand-held portable power tools: measurement of vibrations at the handle — part 7: wrenches, screwdrivers and nut runners with impact, impulse or ratchet action. ISO 8662-7. Geneva: International Organization for Standardization.

International Organization for Standardization (1997). Hand-held portable power tools: measurement of vibrations at the handle — part 8: polishers and rotary, orbital and random orbital sanders. ISO 8662-8. Geneva: International Organization for Standardization.

International Organization for Standardization (1997). Hand-held portable power tools: measurement of vibrations at the handle — part 13: die grinders. ISO 8662-13. Geneva: International Organization for Standardization.

International Organization for Standardization (1997). Mechanical vibration and shock: evaluation of human exposure to whole-body vibration — part 1: general requirements. ISO 2631-1. Geneva: International Organization for Standardization.

International Organization for Standardization (1997). Mechanical vibration and shock: human exposure — biodynamic coordinate systems. ISO 8727. Geneva: International Organization for Standardization.

International Organization for Standardization (1998). Hand-held portable power tools: measurement of vibrations at the handle — part 10: nibblers and shears. ISO 8662-10. Geneva: International Organization for Standardization.

International Organization for Standardization (1998). Mechanical vibration and shock: guidance on safety aspects of tests and experiments with people — exposure to whole-body mechanical vibration and repeated shock. ISO 13090-1. Geneva: International Organization for Standardization.

International Organization for Standardization (1998). Mechanical vibration and shock: hand-arm vibration — method for measuring the vibration transmissibility of resilient materials when loaded by the hand-arm system. ISO 13753. Geneva: International Organization for Standardization.

International Organization for Standardization (1999). Hand-held portable power tools: measurement of vibrations at the handle — part 11: fastener driving tools. ISO 8662-11. Geneva: International Organization for Standardization.

International Organization for Standardization (2000). Earth moving machinery: laboratory evaluation of operator seat vibration. ISO 7096. Geneva: International Organization for Standardization.

International Organization for Standardization (2001). Mechanical vibration and shock: evaluation of human exposure to whole-body vibration — guidelines for the evaluation of the effects of vibration and rotational motion on passengers and crew comfort in fixed-guideway transport systems. ISO 2631-4. Geneva: International Organization for Standardization.

International Organization for Standardization (2001). Mechanical vibration: laboratory method for evaluating vehicle seat vibration — part 2: application to railway vehicles. ISO 10326-2. Geneva: International Organization for Standardization.

International Organization for Standardization (2001). Mechanical vibration: measurement and analysis of whole-body vibration to which passengers and crew are exposed in railway vehicles. ISO 10056. Geneva: International Organization for Standardization.

International Organization for Standardization (2001). Mechanical vibration: measurement and evaluation of human exposure to hand transmitted vibration — part 1: general guidelines. ISO 5349-1. Geneva: International Organization for Standardization.

International Organization for Standardization (2001). Mechanical vibration: measurement and evaluation of human exposure to hand transmitted vibration — part 2: practical guidance for measurement at the workplace. ISO 5349-2. Geneva: International Organization for Standardization.

International Organization for Standardization (2001). Mechanical vibration: vibrotactile perception thresholds for the assessment of nerve dysfunction — part 1: method of measurement at the fingertips. ISO/FDIS 13091-1. Geneva: International Organization for Standardization.

International Organization for Standardization (2001). Mechanical vibration: vibrotactile perception thresholds for the assessment of nerve dysfunction — part 2: reporting and interpretation of measurements at the fingertips. ISO/FDIS 13091-2. Geneva: International Organization for Standardization.

International Organization for Standardization (2002). Agricultural wheeled tractors and field machinery: measurement of whole-body vibration of the operator. ISO 5008. Geneva: International Organization for Standardization.

International Organization for Standardization (2004). Mechanical vibration and shock: cold provocation tests for the assessment of peripheral vascular function — part 1: measurement and evaluation of finger skin temperature. Draft ISO 14835-1. Geneva: International Organization for Standardization.

International Organization for Standardization (2004). Mechanical vibration and shock: cold provocation tests for the assessment of peripheral vascular function — part 2: measurement and evaluation of finger systolic blood pressure. Draft ISO 14835-2. Geneva: International Organization for Standardization.

Japanese Standards Association (1987). Vibration isolation gloves. JIS T 8114. Tokyo: Japanese Standards Association.

Kennedy, R.S., Dunlap, W.P., and Fowlkes, J.E. (1990). Prediction of motion sickness susceptibility. In *Motion and Space Sickness,* Crampton, G.H., Ed. Boca Raton, FL: CRC Press.

Kennedy, R.S., Lilienthal, M.G., Berbaum, K.S., Baltzley, D.R., and McCauley, M.E. (1989). Simulator sickness in U.S. Navy flight simulators. *Aviation Space and Environmental Medicine,* 60(1), 10–16.

Kinsler, L.E., Frey, A.R., Coppens, A.B., and Sanders, J.V. (2000). *Fundamentals of Acoustics.* (4th ed.). Chichester, U.K.: John Wiley & Sons.

Kitazaki, S. (1994). Modelling mechanical responses to human whole-body vibration. Ph.D. thesis. University of Southampton, U.K.

Kjellberg, A. and Wikström, B.O. (1985). Subjective reactions to whole-body vibration of short duration. *Journal of Sound and Vibration,* 99, 415–424.

Kjellberg, A. and Wikström, B.O. (1985). Whole-body vibration: Exposure time and acute effects — a review. *Ergonomics,* 28, 535–544.

Kjellberg, A., Wikström, B.O., and Landström, U. (1994). Injuries and other adverse effects of occupational exposure to whole-body vibration: a review for criteria documentation. *Arbete och Hälsa,* 41.

Kron, M.A. and Ellner, J.J. (1988). Buffer's belly. *The New England Journal of Medicine,* 318(9), 584.

Lalor, N. (1998). Fundamentals of vibration. In *Fundamentals of Noise and Vibration,* Fahy, F.J. and Walker, J.G., Eds. London: E & FN Spon.

Lawther, A. and Griffin, M.J. (1987). Prediction of the incidence of motion sickness from the magnitude, frequency and duration of vertical oscillation. *Journal of the Acoustical Society of America,* 82(3), 957–966.

Lawther, A. and Griffin, M.J. (1988). A survey of the occurrence of motion sickness amongst passengers at sea. *Aviation Space and Environmental Medicine,* 59(5), 399–406.

Lawther, A. and Griffin, M.J. (1988). Motion sickness and motion characteristics of vessels at sea. *Ergonomics,* 31, 1373–1394.

Lerman, Y., Sadovsky, G., Goldberg, E., Kedem, R., Peritz, E., and Pines, A. (1993). Correlates of military tank simulator sickness. *Aviation Space and Environmental Medicine,* 64(7), 619–622.

Lewis, C.H. and Griffin, M.J. (1998). A comparison of evaluations and assessments obtained using alternative standards for predicting the hazards of whole-body vibration and repeated shocks. *Journal of Sound and Vibration,* 215(4), 915–926.

Lewis, C.H. and Griffin, M.J. (2002). Evaluating the vibration isolation of soft seat cushions using an active anthropodynamic dummy. *Journal of Sound and Vibration,* 253(1), 295–311.

Lindsell, C.J. and Griffin, M.J. (1998). Standardised diagnostic methods for assessing components of the hand-arm vibration syndrome. *Health and Safety Executive,* Contract Research Report 197. Sudbury, U.K.: HSE Books.

Lindsell, C.J. and Griffin, M.J. (1999). Thermal thresholds, vibrotactile thresholds and finger systolic blood pressures in dockyard workers exposed to hand-transmitted vibration. *International Archives of Occupational and Environmental Health,* 72(6), 377–386.

Lindsell, C.J. and Griffin, M.J. (2000). A standardised test battery for assessing vascular and neurological components of the hand-arm vibration syndrome. Proceedings of 8th International Conference on Hand-Arm Vibration, 9–12 June 1998, Umeå, Sweden.

Lings, S. and Leboeuf-Yde, C. (2000). Whole-body vibration and low back pain: a systematic, critical review of the epidemiological literature 1992–1999. *International Archives of Occupational and Environmental Health,* 73(5), 290–297.

Lobb, B. (2001). A frequency weighting for motion sickness susceptibility in the lateral axis. Proceedings of the 36th U.K. Group Conference on Human Response to Vibration, held at Centre for Human Science, QinetiQ, Farnborough, U.K., September 12–14, 2001.

Losa, M. and Ristori, C. (2002). Lateral accelerations induced by road alignment: effects on motion sickness and discomfort. Proceedings of 37th U.K. Conference on Human Response to Vibration, held at the Department of Human Sciences, Loughborough University, September 18–20, 2002.

Lundström, R. and Lindberg, L. (1983). Helkroppsvibrationer i entreprenadfordon (Whole-body vibrations in road construction vehicles). The Swedish National Board of Occupational Safety and Health, Investigation Report 1983:18.

Maeda, S. and Griffin, M.J. (1994). A comparison of vibrotactile thresholds on the finger obtained with different equipment. Ergonomics, 37(8), 1391–1406.

Maeda, S., Yonekawa, Y., Kanada, K., and Takahashi, Y. (1999). Whole-body vibration perception thresholds of recumbent subjects — part 2: effect of vibration direction. Industrial Health, 37(4), 404–414.

Magnusson, M.L. and Pope, M.H. (1998). A review of the biomechanics and epidemiology of working postures (it isn't always vibration which is to blame!). Journal of Sound and Vibration, 215(4), 965–976.

Magnusson, M.L., Pope, M.H., Rostedt, M., and Hannson, T. (1993). Effect of backrest inclination on the transmission of vertical vibrations through the lumbar spine. Clinical Biomechanics, 8(1), 5–12.

Mansfield, N.J. (1993). Prediction of the transmissibility of a vehicle seat in the vertical direction. Proceedings of U.K. Informal Group on Human Response to Vibration, Army Personnel Research Establishment, September 20–22, 1993.

Mansfield, N.J. and Griffin, M.J. (1996). Vehicle seat dynamics measured with an anthropodynamic dummy and human subjects. Proceedings of InterNoise 1996, Liverpool, U.K.

Mansfield, N.J. and Griffin, M.J. (2000). Difference thresholds for automobile seat vibration. Applied Ergonomics, 31(3), 255–261.

Mansfield, N.J. and Griffin, M.J. (2000). Non-linearities in apparent mass and transmissibility during exposure to whole-body vertical vibration. Journal of Biomechanics, 33(8), 933–941.

Mansfield, N.J. and Griffin, M.J. (2002). Effects of posture and vibration magnitude on apparent mass and pelvis rotation during exposure to whole-body vertical vibration. Journal of Sound and Vibration, 253(1), 93–107.

Mansfield, N.J., Holmlund, P., and Lundström, R. (1998). Progress in the development of a WBV and HAV database on the Internet. Proceedings of U.K. Informal Group on Human Response to Vibration, Buxton, Derbyshire, September 16–18, 1998.

Mansfield, N.J., Holmlund, P., and Lundström, R. (2000). Comparison of subjective responses to vibration and shock with standard analysis methods and absorbed power. Journal of Sound and Vibration, 230(3), 477–491.

Mansfield, N.J., Holmlund, P., and Lundström, R. (2001). Apparent mass and absorbed power during exposure to whole-body vibration and repeated shocks. Journal of Sound and Vibration, 248(3), 427–440.

Mansfield, N.J. and Lundström, R. (1999a). Models of the apparent mass of the seated human body exposed to horizontal whole-body vibration. Aviation Space and Environmental Medicine, 70(12), 1166–1172.

Mansfield, N.J. and Lundström, R. (1999b). The apparent mass of the human body exposed to non-orthogonal horizontal vibration. Journal of Biomechanics, 32(12), 1269–1278.

Mansfield, N.J. and Marshall, J.M. (2001). Symptoms of musculoskeletal disorders in stage rally drivers and co-drivers. British Journal of Sports Medicine, 35, 314–320.

Matsumoto, Y. and Griffin, M.J. (1998). Dynamic response of the standing human body exposed to vertical vibration: influence of posture and vibration magnitude. *Journal of Sound and Vibration*, 212(1), 85–107.

Matsumoto, Y. and Griffin, M.J. (2002). Effect of muscle tension on non-linearities in the apparent masses of seated subjects exposed to vertical whole-body vibration. *Journal of Sound and Vibration*, 253(1), 77–92.

McCauley, M.E., Royal, J.W., and Wylie, C.D. (1976). Motion sickness incidence: exploratory studies of habituation, pitch and roll, and the refinement of a mathematical model. Report 1733-2, AD-A024 709. Santa Barbara, CA: Human Factors Research, Inc.

McManus, S.J., St.Claire, K.A., Boileau, P.-E., Boutin, J., and Rakheja, S. (2002). Evaluation of vibration and shock attenuation performance of a suspension seat with a semi-active magnetorheological fluid damper. *Journal of Sound and Vibration*, 253(1), 313–327.

Messenger, A.J. (1992). Discomfort and prolonged sitting in static and dynamic conditions. Proceedings of U.K. Informal Group on Human Response to Vibration, Institute of Sound and Vibration Research, Southampton, September 28–30, 1992.

Mills, K.L. and Griffin, M.J. (2000). Effect of seating, vision and direction of horizontal oscillation on motion sickness. *Aviation, Space and Environmental Medicine*, 71(10), 996–1002.

Miwa, T. and Yonekawa, Y. (1971). Evaluation methods for vibration effect, part 10: measurements of vibration attenuation effect of cushions. *Industrial Health*, 9, 81.

Money, K.E. (1990). Motion sickness and evolution. In *Motion and Space Sickness*, Crampton G.H., Ed. Boca Raton, FL: CRC Press.

Money, K.E. and Cheung, B.S. (1983). Another function of the inner ear: facilitation of the emetic response to poisons. *Aviation, Space and Environmental Medicine*, 54(3), 208–211.

Morioka, M. (1999). Effect of contact location on vibration perception thresholds in the glabrous skin of the human hand. Proceedings of the 34th United Kingdom Group Meeting on Human Response to Vibration, held at Ford Motor Company, Dunton, Essex, U.K., September 22–24, 1999.

Morioka, M. (2001). Sensitivity of pacinian and non-pacinian receptors: effect of surround and contact location. Proceedings of the 36th U.K. Group Conference on Human Response to Vibration, held at Centre for Human Science, QinetiQ, Farnborough, U.K., September 12–14, 2001.

Nelson, C.M. (1988). Investigation of the relationship between latent period for vibration-induced white finger and vibration exposure. Proceedings of the U.K. and French joint meeting on Human Response to Vibration, INRS, Vandoeuvre, France, September 26–28, 1988.

Nelson, C.M. (1997). Hand-transmitted vibration assessment: a comparison of results using single axis and triaxial methods. Proceedings of U.K. Informal Group on Human Response to Vibration, Institute of Sound and Vibration Research, Southampton, September 17–19, 1997.

Norheim, A.J., Pedersen, E.J., Fonnebo, V., and Berge, L. (2001). Acupressure treatment of morning sickness in pregnancy: a randomised, double-blind, placebo-controlled study. *Scandinavian Journal of Primary Health Care*, 19(1), 43–47.

O'Hanlon, J.F. and McCauley, M.E. (1974). Motion sickness incidence as a function of the frequency and acceleration of vertical sinusoidal motion. *Aerospace Medicine*, 45, 366–369.

Paddan, G.S. and Griffin, M.J. (1988). The transmission of translational seat vibration to the head, I: vertical seat vibration. *Journal of Biomechanics,* 21(3), 191–198.

Paddan, G.S. and Griffin, M.J. (1988). The transmission of translational seat vibration to the head, II: horizontal seat vibration. *Journal of Biomechanics,* 21(3), 199–206.

Paddan, G.S. and Griffin, M.J. (1993). Transmission of vibration through the human body to the head: a summary of experimental data. ISVR Technical Report 218. Institute of Sound and Vibration Research, University of Southampton.

Paddan, G.S. and Griffin, M.J. (1994). Transmission of roll and pitch seat vibration to the head. *Ergonomics,* 37(9), 1513–1531.

Paddan, G.S. and Griffin, M.J. (1996). Transmission of mechanical vibration through the human body to the head. ISVR Technical Report 260. Institute of Sound and Vibration Research, University of Southampton.

Paddan, G.S. and Griffin, M.J. (1999). Standard tests for the vibration transmissibility of gloves. *Health and Safety Executive,* Contract Research Report 197. Sudbury, U.K.: HSE Books.

Paddan, G.S. and Griffin, M.J. (2002). Effect of seating on exposures to whole-body vibration in vehicles. *Journal of Sound and Vibration,* 253(1), 215–241.

Paddan, G.S. and Griffin, M.J. (2002). Evaluation of whole-body vibration in vehicles. *Journal of Sound and Vibration,* 253(1), 195–213.

Palmer, K.T., Coggon, D., Bednall, H.E., Pannett, B., Griffin, M.J., and Haward, B.M. (1999). Hand-transmitted vibration: occupational exposures and their health effects in Great Britain. *Health and Safety Executive,* Contract Research Report 232. Sudbury, U.K.: HSE Books.

Palmer, K.T., Coggon, D., Bednall, H.E., Pannett, B., Griffin, M.J., and Haward, B.M. (1999). Whole-body vibration: occupational exposures and their health effects in Great Britain. *Health and Safety Executive,* Contract Research Report 233. Sudbury, U.K.: HSE Books.

Palmer, K.T., Griffin, M.J., Syddall, H.E., Pannett, B., Cooper, C., and Coggon, D. (2002). Raynaud's phenomenon, vibration induced white finger and difficulties in hearing. *Occupational and Environmental Medicine,* 59(9), 640–642.

Parrott, A.C. (1989). Transdermal scopolamine: a review of its effects upon motion sickness, psychological performance and physiological functioning. *Aviation, Space and Environmental Medicine,* 60(1), 1–9.

Parsons, K.C. (2003). *Human thermal environments: the effects of hot, moderate and cold environments on human health, comfort and performance.* (2nd ed.). London: Taylor & Francis.

Parsons, K.C. and Griffin, M.J. (1988). Whole-body vibration perception thresholds. *Journal of Sound and Vibration,* 121(2), 237–258.

Peacock, B. and Karwowski, W. (1993). *Automotive ergonomics.* London: Taylor & Francis.

Pelmear, P.L. and Wasserman, D.E. (1998). *Hand-Arm Vibration: A Comprehensive Guide for Occupational Health Professionals.* (2nd ed.). Beverly Farms, MA: OEM Press.

Porter, J.M. and Gyi, D.E. (1998). Exploring the optimum posture for driver comfort. *International Journal of Vehicle Design,* 19(3), 255–267.

Probst, T., Krafczyk, S., Buchele, W., and Brandt, T. (1982). Visuelle pravention der bewegungskrankheit im auto (visual prevention from motion sickness in cars). *Archiv fur Psychiatrie und Nervenkrankheiten,* 231(5), 409–421.

Putcha, L., Berens, K.L., Marshburn, T.H., Ortega, H.J., and Billica, R.D. (1999). Pharmaceutical use by U.S. astronauts on space shuttle missions. *Aviation, Space and Environmental Medicine,* 70(7), 705–708.

Pyykkö, I., Starck, J.P., Korhonen, O.S., Farkkila, M.A., and Aatola, S.A. (1982). Link between noise-induced hearing loss and the vasospastic component of the vibration syndrome. *British Journal of Industrial Medicine,* 45, 188–192.

Rakheja, S., Dong, R., Welcome, D., and Schopper, A.W. (2002). Estimation of tool-specific isolation performance of anti-vibration gloves. *International Journal of Industrial Ergonomics,* 30(20), 71.

Rakheja, S., Haru, I., and Boileau, P.-E. (2002). Seated occupant apparent mass characteristics under automotive postures and vertical vibration. *Journal of Sound and Vibration,* 253(1), 57–75.

Randall, J.M., Duggan, J.A., Alami, M.A., and White, R.P. (1997). Frequency weightings for the aversion of broiler chickens to horizontal and vertical vibration. *Journal of Agricultural Engineering Research,* 68(4), 387–397.

Raynaud, A.G.M. (1862). De l'asphyxie locale et de la gangrène symétrique des extrémités. Ph.D. Thesis, Paris.

Reason, J.T. and Brand, J.J. (1975). *Motion Sickness.* London: Academic Press.

Reynolds, H.M. (1993). Automotive seat design for sitting comfort. In *Automotive Ergonomics,* Peacock, B. and Karwowski, W., Eds. London: Taylor & Francis.

Rubin, C., Recker, R., Cullen, D., Ryaby, J., McCabe, J., and McLeod, K. (2004). Prevention of postmenopausal bone loss by a low-magnitude, high-frequency mechanical stimuli: a clinical trial assessing compliance, efficacy and safety. *Journal of Bone and Mineral Research,* 19(3), 343–351.

Ruffell, C.M. and Griffin, M.J. (2001). Effects of 1-Hz and 2-Hz transient vertical vibration on discomfort. *Journal of the Acoustical Society of America,* 98(4), 2157–2164.

Sanders, M.S., McCormick, E.J. (1993). *Human Factors in Engineering and Design.* New York: McGraw-Hill.

Sandover, J. (1998). High acceleration events: an introduction and review of expert opinion. *Journal of Sound and Vibration,* 215(4), 699–722.

Scarlett, A.J., Price, J.S., and Stayner R.M. (2002). Whole-body vibration: initial evaluation of emissions originating from modern agricultural tractors. *Health and Safety Executive,* Contract Research Report 413. Sudbury, U.K.: HSE Books.

Scarlett, A.J. and Stayner, R.M. (2001). A study of the association between whole-body vibration emission values and daily exposure: case of farm workers. Proceedings of the 36th U.K. Group Conference on Human Response to Vibration, held at Centre for Human Science, QinetiQ, Farnborough, U.K., September 12–14, 2001.

Shen, W. and Parsons, K.C. (1997). Validity and reliability of rating scales for seated pressure discomfort. *International Journal of Industrial Ergonomics,* 20, 441–461.

Shoenberger, R.W. (1972). Human response to whole-body vibration. *Perception and Motor Skills,* 34(1), 127–160.

Shoenberger, R.W. (1982). Discomfort judgements of translational and angular whole-body vibrations. *Aviation, Space, and Environmental Medicine,* 53(5), 454–457.

Shoenberger, R.W. and Harris C.S. (1971). Psychophysical assessment of whole-body vibration. *Human Factors,* 13(1), 41–50.

Smeatham D. (2000). Assessment of vibration from hand-held grinders: the need for a test code based on real operation. Proceedings of U.K. Informal Group on Human Response to Vibration, Institute of Sound and Vibration Research, Southampton, September 13–15, 2000.

Society of Automotive Engineers (1973). Measurement of whole-body vibration of the seated operator of agricultural equipment — SAE J1013.

Society of Automotive Engineers (2003). *SAE Handbook.* Society of Automotive Engineers.

Stainton, M.C. and Neff, E.J.A. (1994). The efficacy of SeaBands for the control of nausea and vomiting in pregnancy. *Health Care for Women International,* 15(6), 563–575.

Stayner, R.M. (1997). European grinder vibration test code: a critical review. *Health and Safety Executive,* Contract Research Report 135. Sudbury, U.K.: HSE Books.

Stayner, R.M. (2001). Whole-body vibration and shock: a literature review: extension of a study of over travel of seat suspensions. *Health and Safety Executive,* Contract Research Report 333. Sudbury, U.K.: HSE Books.

Stevens, S.S. (1957). On the psychophysical law. *Psychological Review,* 64, 153–181.

Stott, J.R.R., Viveash, J.P., and King, S.K. (1993). The transmission of vibration from the hand to the head and its effects on visual stability. Proceedings of U.K. Informal Group on Human Response to Vibration, Army Personnel Research Establishment, September 20–22, 1993.

Swedish Standards Institute. (1989). Handmaskiner: Handhållna motordrivna maskiner — Mätning av vibrationer i handtag — Del 1: Allmänna krav. SS-ISO 8662-1. Stockholm: Swedish Standards Institute.

Tingsgard, I. and Rasmussen, K. (1994). Vibrationsinducerede hvide taeer (Vibration-induced white toes). *Ugeskr Laeger,* 156(34), 4836–4838.

Triesman, M. (1977). Motion sickness: an evolutionary hypothesis. *Science,* 197, 493–495.

Turner, M. and Griffin, M.J. (1995). Motion sickness incidence during a round-the-world yacht race. *Aviation, Space and Environmental Medicine,* 66(9), 849–856.

Turner, M. and Griffin, M.J. (1999). Motion sickness in public road transport: passenger behaviour and susceptibility. *Ergonomics,* 42(3), 444–461.

Turner, M. and Griffin, M.J. (1999). Motion sickness in public road transport: the effect of driver, route and vehicle. *Ergonomics,* 42(12), 1646–1664.

Turner, M., Griffin, M.J., and Holland, I. (2000). Airsickness and aircraft motion during short-haul flights. *Aviation, Space and Environmental Medicine,* 71(12), 1181–1189.

Verschueren, S.M.P., Roelants, M., Delecluse, C., Swinnen, S., Vanderschueren, D., and Boonen, S. (2004). Effect of 6-month whole body vibration training on hip density, muscle strength and postural control in postmenopausal women: a randomized controlled pilot study. *Journal of Bone and Mineral Research,* 19(3), 352–359.

Videman, T., Simonen, R., Usenius, J.P., Österman, K., and Battié, M.C. (2000). The long-term effects of rally driving on spinal pathology. *Clinical Biomechanics,* 15(2), 83–86.

Ward, T. (1996). Correlation between vibration emission data and vibration during use: grinders. Health and Safety Laboratories, *Research Report IR/L/NV/96/20.*

Ward, T. (1996). Hand-tool vibration emission: correlation between declared and in-use values: part 1 — grinders. Proceedings of U.K. Group Meeting on Human Response to Vibration, MIRA, September 18–20, 1996.

Warwick-Evans, L.A., Masters, I.J., and Redstone, S.B. (1991). A double-blind placebo controlled evaluation of acupressure in the treatment of motion sickness. *Aviation, Space and Environmental Medicine,* 62(8), 776–778.

Wei, L., Griffin, M.J. (1998). Mathematical models for the apparent mass of the seated human body exposed to vertical vibration. *Journal of Sound and Vibration,* 212(5), 855–874.

Whitehouse, D.J. (2002). The effect of probe size on vibrotactile thresholds at the fingertip and volar forearm. Proceedings of the 38th U.K. Conference on Human Response to Vibration, held at Institute of Naval Medicine, Alverstoke, Gosport, PO12 2DL, September 17–19, 2003.

Wikström, B.O., Kjellberg, A., and Dallner, M. (1991). Whole-body vibration: a comparison of different methods for the evaluation of mechanical shocks. *International Journal of Industrial Ergonomics,* 7, 41–52.

Wu, X. and Griffin, M.J. (1996). Towards the standardization of a testing method for the end-stop impacts of suspension seats. *Journal of Sound and Vibration,* 192(1), 307–319.

Wu, X. and Griffin, M.J. (1998). The influence of end-stop buffer characteristics on the severity of suspension seat end-stop impacts. *Journal of Sound and Vibration,* 215(4), 989–996.

Yonekawa, Y., Maeda, S., Kanada, K., and Takahashi, Y. (1999). Whole-body vibration perception thresholds of recumbent subjects: part 1 — supine posture. *Industrial Health,* 37(4), 398–403.

Index

A

A(8), 83, 117–118, 170–173, 175, 193
Acceleration, 2–5
Accelerometers, 98, 104–106
Action values, *see* Exposure action values
Active control, 40, 51
Activity interference, 8–9, 23
ADC, *see* Analog-to-digital converter
Adding exposures, 116–118
Adding waves, 2–4
Agricultural tractors, *see* Tractors
Agriculture, 197
Aircraft, 16, 63, 75
Aircraft simulators, 65, 66
Airsickness, 63–64, 75
Alcohol, 58–60
Aliasing, 103–104
Alternative therapies, 71–72
Ambulances, 7, 16–17
Amplifiers, 106–108
Amplitude, 2–3, 8–9
Analog-to-digital converter, 101–102
Anger time, *see* Trigger time
Animal welfare, 19
Anthropodynamic dummies, 41, 45
Antialiasing filters, 104, 109, *see also* Aliasing
Anticholinergic drugs, 72–73
Antihistamine drugs, 72–73
Antivibration gloves, 94–95, 181–185
Apparent mass, 44–47
Armored fighting vehicles, 65; *see also* Tanks
Artifacts, 99, 109, 129
Astronauts, 73
Auditory system, 14–15; *see also* Hearing
Automobiles, *see* Cars
Avalanche phenomenon, 55
Avoidance, 71, 74–75
Axes, *see* Vibration axes; Direction
Axis multipliers, *see* Multiplication factors
Axis summation, 115–116

B

Babies, 75
Backhoe loaders, 29

Back pain, 23–31, 85, 196, 197
Backrest, 43, 148, 150, 151–152
Biodynamics, 42; *see also* Biomechanics
Biomechanics, 1–2, 10, 41–52
Blanching, 80, 81, 86–88, 92
Boats, 55, 58, 60–63; *see also* Ships
Bone and joint disorders, 84
Bone density, 84
Brammer's scale, 87
Bridges, 5
British Standards
    BS 1032, 158–160
    BS 1033, 175–176
    BS 5008, 155–157
    BS 5349, 170–173
    BS 6472, 20
    BS 6611, 20
    BS 6841, 16–17, 19, 62–63, 114, 147–150,
        153–155
    BS 6842, 169, 173–174
    BS 6912, 164
    BS 7096, 164
    BS 8662, 179–181
    BS 10056, 157
    BS 10326, 162–164
    BS 10819, 183–185
    BS 13059, 160
    BS 13490, 164–166
    BS 13753, 182–183
    BS 28662, 176–177
    BS 30326, 162–163
British Standards Institute, 145–146
BSI, *see* British Standards Institute
Buffer's belly, 85
Buffers, *see* End-stop buffers
Building vibration, 4, 18, 19–20
Buses, 16, 56, 64–65, 71, 122

C

Calibration, 109–110, 123
Calibrators, 109
Camels, 53
Car industry, 8, 21
Carpal tunnel syndrome, 84
Cars, 8, 14, 25, 27–28